D1217952

Fish Vaccination

Plate 1 Rainbow trout have been immersed in a tank containing the diluted vaccine for 30 s. The fish are then removed and the vaccine is allowed to drain.

Plate 2 The dip vaccinated fish are placed into an adjacent tank.

Photographs, courtesy of Dr. M. T. Horne, Stirling Diagnostics Ltd.

Fish Vaccination

Edited by

Dr A E Ellis

DAFS
Marine Laboratory
PO Box 101
Victoria Road
Aberdeen AB9 8DB

1988

ACADEMIC PRESS
Harcourt Brace Jovanovich, Publishers
London San Diego New York Berkeley
Boston Sydney Tokyo Toronto

ACADEMIC PRESS LIMITED
24/28 OVAL ROAD
LONDON NW1 7DX

United States Edition published by
ACADEMIC PRESS INC.
San Diego CA 92101

British Library Cataloguing in Publication Data

Ellis, A. E.
Fish Vaccination
1. Fish. Vaccination
I. title
597′.029

ISBN 0-12-237485-1

Typeset by Bath Typesetting Ltd, Bath
Printed in Great Britain by St Edmundsbury Press, Bury St Edmunds, Suffolk

Contributors

D. W. BRUNO, *Marine Laboratory, Torry, Aberdeen, Scotland, UK*

M. DORSON, *Laboratoire d'Ichthyopathologie, INRA, Route de Thiverval-Grignon, 78850 Thiverval-Grignon, France*

A. E. ELLIS, *Department of Agriculture and Fisheries for Scotland, Marine Laboratory, PO Box 101, Victoria Road, Aberdeen AB9 8DB, Scotland, UK*

N. FIJAN, *Department for Biology and Pathology of Fish and Bees, Veterinary Faculty, University of Zagreb, Zagreb, Yugoslavia*

J. L. FRYER, *Department of Microbiology, Oregon State University, Corvallis, OR 97331–3804, USA*

J. E. HARRIS, *Department of Biological Sciences, Plymouth Polytechnic, Plymouth, Devon, UK*

T. S. HASTINGS, *Marine Laboratory, PO Box 101, Victoria Road, Aberdeen AB9 8DB, Scotland, UK*

M. T. HORNE, *Stirling Diagnostics Limited, Innovation Park, Hillfoot Road, Stirling FK9 4NF, Scotland, UK*

G. HOUGHTON, *Department of Biological Sciences, Plymouth Polytechnic, Plymouth, Devon, UK*

P. DE KINKELIN, *Laboratoire d'Ichthyopathologie, INRA, Route de Thiverval-Grignon, 78850 Thiverval-Grignon, France*

J. C. LEONG, *Department of Microbiology, Oregon State University, Corvallis, OR 97331–3804, USA*

R. A. MATTHEWS, *Department of Biological Sciences, Plymouth Polytechnic, Plymouth, Devon, UK*

A. L. S. MUNRO, *Marine Laboratory, Torry, Aberdeen, Scotland, UK*

J. A. PLUMB, *Department of Fisheries and Allied Aquacultures, Alabama Agricultural Experiment Station, Auburn University, Alabama 36849, USA*

F. SALATI, *Via Doni 5, Roa Piana, 12080 Monastero di Vasco (CN), Italy*

C. J. SECOMBES, *Department of Zoology, University of Aberdeen, Tillydrone Avenue, Aberdeen AB9 2TN, Scotland, UK*

P. D. SMITH, *Aquaculture Vaccines Limited, 24–26 Gold Street, Saffron Walden, Essex CB10 1EJ, England, UK*

R. M. W. STEVENSON, *Department of Microbiology, College of Biological Science, University of Guelph, Guelph, Ontario N1G 2W1, Canada*

J. R. WINTON, *National Fisheries Research Center, Bldg. 204, Naval Support Activity, Seattle WA 98115, USA*

Preface

Over the last 20 years, aquaculture has grown into a very significant industry in many parts of the world. In most developed countries fish are farmed intensively and under such conditions of high population density, infectious diseases pose a constant and highly costly threat to successful animal husbandry. Even when environmental conditions are good and fish are healthy certain infectious agents, if introduced into the farm, are so virulent that mass mortalities can and do occur. Antibiotics provide a useful means of helping to control many bacterial diseases but there are many problems associated with the development of antibiotic resistance and recurrent outbreaks necessitating further, costly, treatments. There is no effective chemotherapy for control of viral diseases. It is against this background that vaccines have been perceived as potentially playing an important role in aquaculture.

This perception was given an enormous boost in the mid-1970s with the development of highly effective vaccines against vibriosis and enteric red-mouth (ERM) which could be mass administered by simply dipping or bathing fish in a relatively simply prepared killed broth culture of the pathogens. These vaccines are cheap to produce, easy to administer, and provide long-lasting immunity. In a flurry of activity, similar vaccines against all the other major diseases of cultured fish were subsequently prepared and tested. None has proved consistently effective. However, some degree of protection has been achieved and many researchers are highly optimistic that reliable effective vaccines can be developed to most infectious diseases of fish. Often, the impression given has been too optimistic since many trials have not been designed to test specificity or longevity of protection. There is a strong probability that short-term protection may be due to stimulation of non-specific defence mechanisms by certain components of vaccines. Such protection has limited value in aquaculture.

Thus, the development of many vaccines for fish is still at the experimental stage but it is clear that the simple empirical approach has not been successful other than with the vibriosis and ERM vaccines. Planned strategies for developing vaccines by scientific investigations are now the pattern of research. There are three main stages to this approach.

(i) The identification of protective antigens i.e. those antigens which are capable of stimulating a protective immune response in fish. Hence the importance of analysing the antigenic components of pathogens especially in identifying those antigens which are involved in virulence i.e. the molecules which permit the pathogen to attach, multiply and cause damage in the fish. It is likely that an immune response which neutralizes the effects of the virulence factors will provide effective immunity.

(ii) Once identified, methods of producing the protective antigens in culture must be established.

(iii) Methods of delivering the protective antigens in an immunogenic form must be developed. This may mean some form of concentration of the antigen and even modification to ensure potent stimulation of the immune response and long-lasting protection.

Thus, the development of further vaccines for fish depends upon detailed knowledge of aspects of the biochemistry of the pathogen and the host. In no less measure, the effective deployment of a vaccine depends upon knowledge of the biology of the pathogen and the host i.e. the epidemiological factors of a disease. This book attempts to embody what is currently known in these fields of knowledge, to highlight areas of ignorance, and to assess the effectiveness to which knowledge has been applied to vaccinating fish against most of the major infectious diseases affecting the aquaculture industry.

The first five chapters deal with the principles of vaccination; factors affecting the immune responsiveness of fish; maturation of the immune system in young fish; commercial aspects of vaccine production and licensing; and the strategic use of vaccines on farms. Following these are chapters dealing with vaccination against individual diseases caused by bacteria and viruses. I have attempted to cast these chapters into a set format so that each addresses issues relevant to the development, composition, testing and strategic use of vaccines. Finally, there are two chapters which discuss the recent interest in vaccinating fish against parasite diseases and the potential for immunizing fish in order to manipulate certain physiological processes especially the development of sexual maturation.

I hope the book, by its format and content will be equally useful to both the research worker involved in developing vaccines for fish and also to the fish farmer who, already accustomed to using sophisticated technology as an aid to good husbandry, will need to make complex decisions on how, when and where to vaccinate the stock.

Contents

1

General Principles of Fish Vaccination

A. E. Ellis

WHAT IS A VACCINE

Exposure of a vertebrate animal to an infection often results in survivors becoming resistant to subsequent disease caused by the same pathogenic organism. This resistance, called adaptive immunity, is specific to the challenging pathogen and persists for a relatively long period of time (immune memory). It is based upon an adaptive change in the body's lymphocyte populations resulting from exposure to the foreign molecules constituting the pathogen. These foreign molecules are called antigens and are nearly always proteins. Specificity and memory are two of the key elements exploited by vaccination since the adaptive immune response is stronger on second encounter with the antigen. Vaccines are preparations of antigens derived from pathogenic organisms, rendered non-pathogenic by various means, which will stimulate the immune system in such a way as to increase the resistance to disease from subsequent infection by a pathogen.

Inducing a protective immune response to a pathogenic organism before the individual becomes naturally exposed to it (prophylaxis) seems at first sight an eminently sensible way of preventing an infectious disease. Unfortunately, achieving this goal is by no means easy. Vaccines must be safe (no side effects) and potent (induce a high level of protection) and although occasionally it has been possible to achieve these characteristics using simple procedures (usually heat or chemical inactivation of cultures of pathogenic micro-organisms) in most cases some degree of antigen purification or enrichment is necessary and the problem lies in identifying the relevant antigens, out of many hundreds present, which are important in stimulating a protective immune response.

Vaccines are generally of two types: *dead vaccines*, which are composed of inactivated pathogens or extracts, and *live vaccines*, which are attenuated pathogens with no or low virulence. However, modern techniques have substantially improved and diversified methods of vaccine production.

Before describing what vaccines do and how they may be used it is necessary first to understand the basis of how they work by stimulating the specific immune system.

FISH VACCINATION
ISBN 0-12-237485-1

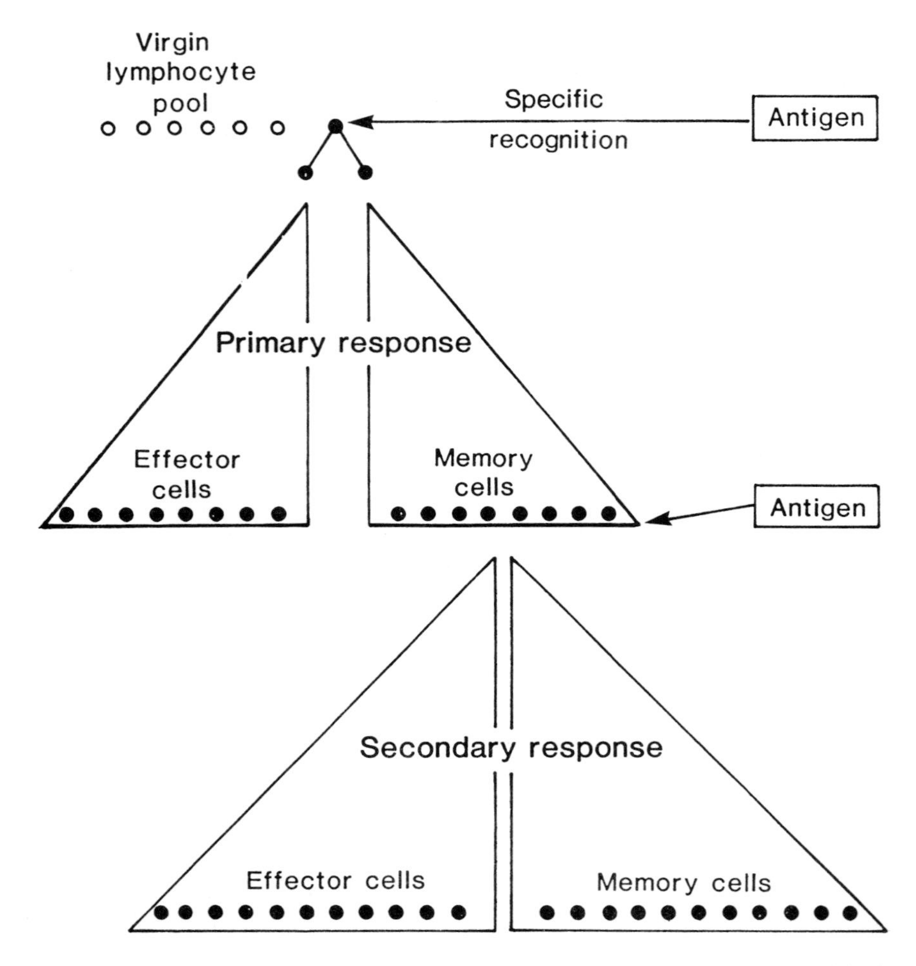

Fig. 1.1 T and B lymphocytes carrying specific antigen recognition sites are produced in the primary lymphoid organs and comprise the virgin lymphocyte pool. On stimulation by antigen a specific clone proliferates and matures into effector cells (B cells become plasma cells producing antibody; T cells become cytotoxic or other effector cells) or memory cells. This constitutes the *primary response*. On subsequent stimulation by antigen, the memory cells again proliferate into effector and memory cells the numbers of which are then greatly expanded accounting for the stronger *secondary response*.

THE SPECIFIC IMMUNE SYSTEM

The specific or adaptive immune response has two arms: humoral immunity (antibody production) and cell mediated immunity (CMI). Exposure to an antigen results in the stimulation of a small number of virgin lymphocytes which are capable of recognizing the antigen through specific antigen receptors (Fig. 1.1). These specific lymphocytes constitute a clone. Micro-organisms have many different antigens on their surface and each antigen is capable of being recognized by a different clone of lymphocytes. On stimulation a clone undergoes proliferation with the differentiation of daughter cells which have particular functions depending upon the population to which the clone of lymphocytes belongs.

There are two main populations of lymphocytes. The 'T' lymphocytes which originate from the thymus, and the 'B' lymphocytes which originate from the bursa of Fabricius in birds or the bone marrow of mammals. The source of this cell type in teleost fish is not definitely known though it is likely to be the kidney.

Humoral immunity

On primary exposure to an antigen T and B cells cooperate in the response. B lymphocytes differentiate into plasma cells which produce antibody specific to the stimulating antigen or into cells which are capable of becoming plasma cells on subsequent exposure to antigen and are therefore called memory cells.

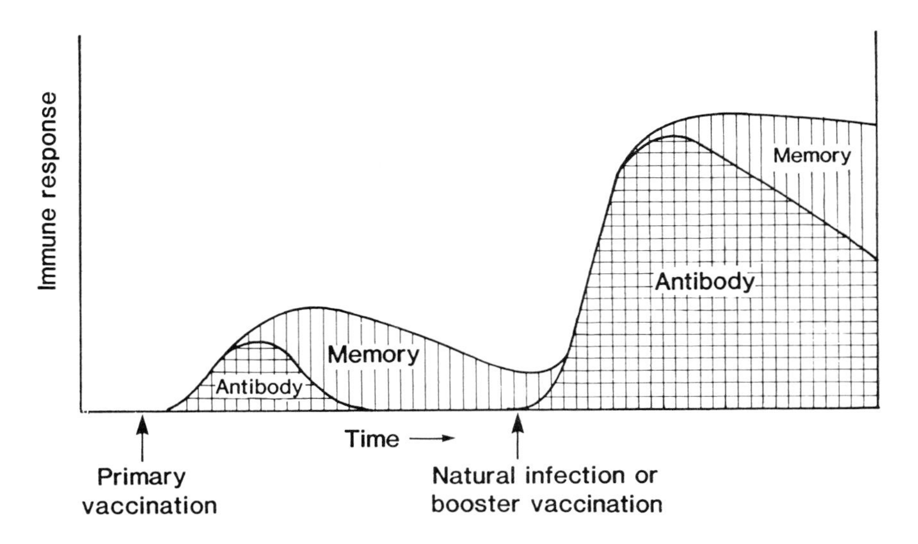

Fig. 1.2 Relationship between magnitude of the antibody response on primary and secondary immunization and the establishment and persistence of immune memory.

The T cells have a different function. The original cooperating T cell is termed a helper cell and this clone, on original stimulation by antigen, proliferates into long-lived helper memory cells so that on subsequent exposure to antigen there will be an increased number of T helper cells to cooperate with the increased number of B memory cells so that antibody production to an antigen on secondary exposure appears in the blood more rapidly and reaches higher concentration than after primary exposure (Fig. 1.2).

It is this ability of the memory cells to effect a rapid and high level response that accounts for the increased resistance following exposure to a pathogen or a vaccine.

Cell-mediated immunity (CMI)

The T lymphocyte population besides possessing T-helper clones also possesses clones which are responsible for CMI. This aspect of the immune response is wide-ranging and recruits the macrophages which constitute the body's main line of non-specific defence by phagocytosing and digesting invading micro-organisms. On primary antigen stimulation the T lymphocyte clones differentiate into several different functional cell types involved in CMI. These include:

(a) *Killer cells.* These cytotoxic T cells are capable of lysing foreign cells by a mechanism depending upon physical contact between the T cell and the target cell.

(b) *Lymphokine producing cells.* Upon antigen stimulation a clone of responsive T cells release humoral factors called lymphokines (or interleukins) which enhance the non-specific defence capacity of macrophages. These 'activated' macrophages are then much more capable of fighting off infection non-specifically.

(c) *Suppressor cells.* The above responses are termed positive immunity. Obviously, the production of antibodies and lymphokines must be regulated during the course of an immune response and this is performed by the proliferation of suppressor cells which provide the switching off process or negative immunity. This part of the specific immune response is highly relevant to vaccination since the outcome of administering antigen, depending upon a variety of factors including its molecular form, route of administration, concentration or age of the animal, may affect the balance between positive and negative immunity. Obviously the purpose of vaccination is to stimulate the positive aspect preferentially but it is possible to induce the opposite effect and instead of inducing resistance, treatment with antigen under certain circumstances could lead to unresponsiveness to antigen, termed immune tolerance (see also Ch. 3).

Recent work with mammals has shown that antigenic molecules possess different recognition sites (epitopes) for the B cells, T helper cells (Th) and T suppressor (Ts) cells (Fig. 1.3). The relative immunogenicity (capacity to stimulate an immune response) of an antigen in an animal depends upon the relative dominance of these epitopes. This is genetically determined, and can differ even between strains of mice. It has proved possible to enhance the immunogenicity of an antigen by enzymatically removing the epitope responsible for stimulating the Ts cells.[1] This form of protein engineering may prove extremely important in the future in producing highly effective vaccines.

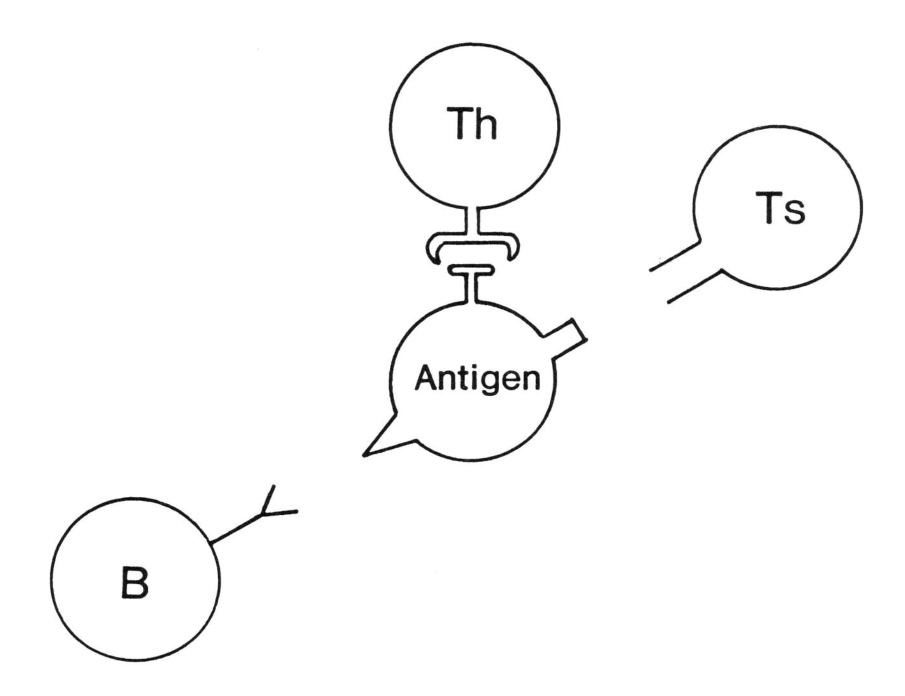

Fig. 1.3 Different functional populations of lymphocytes recognize different epitopes on an antigen (B = B lymphocyte; Th = T-helper cell; Ts = T suppressor cell). In certain circumstances the immunogenicity of an antigen may be enhanced by removal of the epitope(s) recognized by the Ts cell.

WHAT DOES A VACCINE DO?

The purpose of vaccination is to provide an individual with resistance to a disease without having to undergo a potentially risky infection. It is important, however, to appreciate that the protection which develops following vaccination or natural exposure to a pathogen is protection against disease, not necessarily infection. Indeed, the nature of the parasitic relationship between host and pathogen should not be regarded as a lethal struggle between the antagonists. For well-adapted parasites the nature of the relationship is one of co-existence with the minimum damage to the host. Of the 150 or so disease organisms which affect humans many are known to occur in an asymptomatic carrier state in the absence of or following disease signs.[2] A similar condition also applies to fish and many of their pathogens. It is reasonable to argue that the immune response is instrumental in smoothing the path to this peaceful co-existence despite, in some instances, a stormy start to the association between host and parasite with often considerable mortality of both. Thus, vaccines protect against outbreaks of disease but not necessarily against persistent infection in an asymptomatic carrier state.

Vaccines work by inducing a protective immune response which, by virtue of the memory cells, persists for relatively long periods of time, though the precise longevity varies. It is the establishment of this memory state to which the term ‘immunization’ usually refers. Natural exposure to infection acts as a booster to the immunity produced by vaccination provided the memory cells are still present (Fig. 1.2). In the absence of natural exposure, booster vaccination is necessary to maintain the level of immunity.

THE SITES OF PROTECTIVE IMMUNITY

The two arms of the adaptive immune response may act either systemically within the tissues of the body, especially the organs of phagocytic filtration (spleen and kidney of teleost fish) or in integumentary organs (gut, gills and skin). To a certain extent the systemic and integumentary systems respond independently of each other. It can be important therefore that administration of a vaccine be performed in such a way as to stimulate the most appropriate immune response to provide protection. In veterinary medicine vaccines may be administered by injection, orally or by nasal sprays in order to stimulate an immune response in an appropriate place. In fish, little is known about integumentary immune responses but as immersion vaccination against vibriosis, while being protective, results in very poor systemic antibody production, it is likely that protective immunity is based in the mucous membranes. It is possible that protection against some other diseases of fish requires a strong systemic response and the current practice of immersion vaccination may be inadequate to stimulate this. In such cases injection of vaccine may be unavoidable unless improved methods of immersion or oral vaccination are developed.

WHAT MAKES A VACCINE EFFECTIVE?

An effective vaccine must be '*safe*', *immunogenic* and *protective*.

Safety

A vaccine must simulate the natural infection in producing immunity but it must not produce clinical illness. Killed or inactivated vaccines are usually safer than live ones but even these can produce unwanted side effects. Living vaccines require special caution. They are usually attenuated but they are not avirulent. Before they can be used safely it is essential to prove beyond reasonable doubt that an attenuated strain will not mutate to one of enhanced or full virulence *in vivo*. De Kinkelin (Chapter 14) describes some of the problems encountered with live VHS virus vaccines. Modern techniques of genetic engineering have made the process of attenuation more precise and irreversible and rapid progress is taking place in the development of more effective and safe live vaccines for use in human medicine.

Immunogenicity

Not all the antigens associated with virulence and pathogenicity of a microbial pathogen are effective stimulators of the immune response. This appears to be the case with many of the antigens associated with *Aeromonas salmonicida* which are very poorly immunogenic in the salmonid host yet are immunogenic in rabbits. Poor immunogenicity in the host, especially of antigens associated with virulence, may represent an adaptation of the pathogen to avoid destruction by the host's immune system. In such cases it is necessary to alter the antigen to improve immunogenicity in the host but still retain antigenic sites which induce antibodies to the native molecule. Methods of doing this include attachment to immunogenic carriers e.g. proteins; adsorption to adjuvants e.g. alum; or cleaving the molecule to remove epitopes which stimulate the T suppressor cells.

Stimulation of protective immunity

Obviously the purpose of vaccination is to protect against disease and so a vaccine must include protective antigens i.e. those which contribute to the production of disease and which stimulate the production of protective antibodies or CMI when used in immunization. To be of value, therefore, a vaccine must induce an immune response to virulence factors of the pathogen. Hence the importance of work into virulence mechanisms.

All micro-organisms and their products are composed of many different antigens and while many will induce antibody production, unless these antibodies can neutralize the components of pathogenicity they are valueless in giving protection. It is important to realize then that assays of antibody levels in the blood serum following vaccination may give no indication of the degree of protection unless they are against protective antigens. A good example is the induction of high levels of agglutinating antibodies in salmonids vaccinated with *A. salmonicida* bacterins but these antibodies are not at all protective.

Conventional methods of preparing vaccines involve the growth of pathogens in artificial culture media and viruses in tissue culture. Under these conditions antigenic changes may occur such that factors of importance *in vivo* are not produced. Hence *in vitro* cultures are often deficient in protective antigens. This applies more to dead or inactivated vaccines than live ones, but even attenuated strains or strains of close antigenic similarity to the virulent pathogen may be partially inadequate e.g. attenuated strains of IPN virus. The application of knowledge of the special conditions required to produce protective antigens using *in vitro* culture will obviously improve the efficacy of vaccines.

VACCINATION AND OTHER DISEASE CONTROL METHODS

Besides vaccination there are other major ways of controlling animal diseases: exclusion of the pathogen and chemotherapy.

Exclusion

In areas where a disease is not endemic it is possible to exclude the causative agents by a legislative policy. Many countries operate such policies where the importation and movement of live fish or eggs must be accompanied by disease certification. If this occasionally fails and an outbreak of a disease occurs within a 'disease-free zone', spread of the disease can be controlled by slaughter of the stock and disinfection of the premises. In the absence of government compensation for such policies, the fish farmers must be willing to cooperate voluntarily. Nevertheless, there are several examples where slaughter/disinfection policies have worked well e.g. control of VHS in Brittany and parts of Denmark.

Exclusion methods may be far reaching and include pasteurization of trash fish used as food for salmonids to avoid outbreaks of vibriosis; provision of safe water supplies e.g. artesian well water often used in hatcheries; control of disease vectors e.g. freshwater snails which are an intermediate host of *Diplostomum* (eye-fluke), or feral fish species which may act as reservoirs of infectious agents.

Chemotherapy

Treatment of established infections with antimicrobial compounds has played and will continue to play a very important role in the control of many infectious diseases. The one area they have not proved useful is the treatment of viral diseases. Unfortunately, the treatment of bacterial diseases with antimicrobials suffers from the appearance of drug resistant strains and antimicrobial sensitivity tests must be conducted before treatment can commence to ensure the use of an effective drug. Obviously this takes some time (at least 48 h) during which the infection spreads even further in a farmed population. The most practical means of administering a drug is incorporation into the food but this is not ideal since infected fish usually lose appetite and therefore receive inadequate treatment.

Furthermore, there is concern about the effects on the environment of the use of large quantities of chemotherapeutants especially the potential risks of antimicrobial resistance acquired by animal bacterial pathogens being transferred to human pathogens. It is for these reasons that licensing authorities generally impose strict limitations on the use of chemotherapeutants.

Though not generally practised in aquaculture, and with the attendant risk of selecting resistant strains, chemotherapy can be used prophylactically where fish may be at risk from bacterial infection over a short time period or when transferred to a totally new environment. It has been used to treat Atlantic salmon smolts before transfer to seawater in areas where furunculosis outbreaks were expected.

Vaccination – its use on controlling disease

In areas where a disease is not endemic a policy of exclusion and eradication is the best means of control. However, such policy requires exceptionally high levels of vigilance, certification and cooperation between farmers and responsible authorities. On occasions it may fail and the introduction of a disease into a water course may result in it becoming endemic. Many diseases of fish are already endemic in wild fish populations which act as an unavoidable reservoir of infection for farms on surface water supplies. In these areas the risk of farmed stock becoming infected may be high and a slaughter/disinfection policy becomes extremely expensive and probably ineffective. For many of the major diseases of fish (e.g. furunculosis, BKD, VHS) survivors of an outbreak become carriers which restricts the marketing potential of live stock (i.e. transfer to non-infected sites is prohibited) and every new year class of fish is at risk. For bacterial diseases, outbreaks can be controlled by chemotherapy but this may fail for reasons outlined above. In such cases, when a disease is endemic, vaccination becomes useful, especially with killed vaccines. One perceived problem with vaccination is the risk that vaccinated stock, while being resistant to disease, may become carriers of a pathogen after exposure. This is no more likely to occur than with non-vaccinated survivors of natural infection but unless the carrier status of vaccinated populations is closely and accurately monitored, the transfer of "healthy" though carrier fish could become a hazard in itself by spreading the disease to susceptible populations. Even with the potential hazards of live vaccines, there is a case for their use in high risk areas where a virus disease is endemic. This case is argued by de Kinkelin for live VHS vaccines (Chapter 11). These vaccines are slightly virulent themselves and may cause low levels of mortalities with the population becoming carriers of the vaccinating strains. Even so, this scenario is better than high levels of mortalities from VHS with survivors becoming carriers of the virulent wild type virus.

Thus, vaccination has an important role to play in disease control but its use should depend upon various epidemiological considerations of each disease and the nature of the vaccine. Presently only two vaccines have been proven effective on a commercial scale; vibriosis and enteric redmouth vaccines. Many others are at an experimental stage.

HISTORY OF FISH VACCINATION

The first attempts to vaccinate fish against disease were published by Duff[3] who attempted to orally vaccinate cut-throat trout against furunculosis with a diet containing chloroform-killed *Aeromonas salmonicida* cells. Upon a water-borne challenge vaccinated fish showed good protection (25 per cent mortality compared with 75 per cent in non-vaccinated controls). Despite many trials with furunculosis vaccines, Duff's original success has not been reliably reproduced even using other techniques of vaccine administration and antigen preparations. While furunculosis vaccines are marketed commercially, present information suggests their efficacy is marginal.

The first commercial product licence for a vaccine for fish was granted in the USA in 1976 for use against enteric redmouth (ERM). Since then product licences have been granted for *Vibrio* vaccines in many parts of the world and these two vaccines have proved of great commercial value.

The history of the development of mass administration of vaccines by the immersion method which has been so successful with *Vibrio* and ERM vaccines is described in Chapter 6. Many types of fish farming are highly intensive and involve the rearing of very large numbers of fish. Vaccination by injection is therefore not a convenient method though it is not entirely impractical for fish over 15 g. However, mass vaccination techniques are obviously preferred and are essential for very small fish.

Various techniques of mass vaccination of fish have been developed and these include immersion, hyperosmotic immersion, bath, spray and oral.

COMPARISON OF VACCINATION METHODS

Injection vaccination

The most effective way of immunizing fish is by injection intraperitoneally. Furthermore, this method permits the use of adjuvants which enhance the magnitude of the immune response (*adjuvare* is Latin meaning 'to help'; see Ch. 3). The method suffers from certain disadvantages, principally by requiring anaesthetization and handling, both of which can be quite stressful for the fish. It is also very labour intensive and cannot be used with fish much below 15 g. However, by using repeater syringes and a production line system it is quite feasible for each operator to inject 1000 fish/h. Basically the system consists of 'vaccinator(s)' seated at a vaccination table, a method of returning fish to the holding facility (a chute, conveyor belt) and an 'anaesthetist' controlling the supply of anaesthetized fish to the vaccination table for immediate injection thus preventing too great an exposure to the anaesthetic. This system is illustrated in Fig. 1.4.

Oral vaccination

The protocol of oral vaccination is very attractive since it is suitable for mass administration to fish of all sizes, imposes no stress on the fish because handling is not required and therefore does not interfere with routine husbandry practices. Furthermore, oral vaccination is the only method suitable for extensive pond rearing of fish where catching the fish prior to harvesting for injection or immersion vaccination is impractical. However, oral vaccination has some intrinsic limitations. In the case of killed vaccines large amounts are required thus increasing the cost and there is the uncertainty of individual dosage since fish consume different amounts of food. Unfortunately, the most important limitation of oral vaccines is their poor potency and many trials with many different vaccines have resulted in low or inconsistent levels of protection.

Most work on oral administration has been done with *Vibrio* vaccines and in Chapter 6 Smith outlines some of the reasons for failure and possible ways of overcoming them which are applicable to oral vaccines generally.

A major problem seems to be the destruction of antigens in the stomach and foregut before they reach the immune sensitive areas of the lower gut.[4] Future developments will need to overcome this problem, possibly by encapsulating the vaccine so that it survives passage through the stomach. There are few reports of oral vaccine trials in fish which lack a stomach, like the cyprinids, and the difficulties may be less or different for such species.

Other relevant research will be the development of suitable adjuvants for oral vaccines (see Ch. 3). Smith further argues that while present oral vaccines cannot be recommended for primary vaccination against vibriosis they may be useful as booster vaccines. Indeed, little work has been carried out to assess the value of combining different vaccination methods where oral delivery might be used as a priming or boosting method in association with one of the immersion methods.

Fig. 1.4a Injection vaccination. A, large holding tank, B, anaesthetic tank. B1, safe tank with low dose anaesthetic. C, Injection station (Courtesy Dr D. Bruno, Marine Laboratory, Aberdeen).

Fig. 1.4b Injection (intraperitoneal) using repeater syringes attached to vaccine reservoir (Courtesy Dr D. Bruno, Marine Laboratory, Aberdeen).

Immersion vaccination

The several variations of immersion vaccination were developed mainly by Amend and colleagues in the USA for use of the *Vibrio* vaccines (see Ch. 6). The method began with hyperosmotic immersion (HI)[5,6] but it was quickly realized that the hyperosmotic step (which was quite stressful) was unnecessary and direct immersion (DI), in the vaccine was equally successful.[7] The method is simple and rapid, only a few seconds exposure to the vaccine are necessary. This method can now be automated by use of immersion vaccination machines (see Fig. 1.5 and Ch. 6). The 'bath' or 'flush' variation was developed for use of the *Vibrio* and ERM vaccines and simply involves pouring the vaccine into the holding tanks thereby avoiding the handling stress of DI. This method consumes more vaccine and requires longer exposures (about 1 h) which involves oxygenation of the water and close monitoring of fish for signs of stress.

Fig. 1.5 Direct immersion vaccination. The vaccine is diluted in water and fish are immersed for 30–60 s. (Courtesy G. Rae, Marine Harvest Ltd.)

A further variation is the spray method, again developed for use of *Vibrio* vaccines, but has no advantages over DI and is more stressful to the fish.

While immersion (DI) vaccination has proved very efficacious for immunizing fish against vibriosis and ERM it has not been so successful for other diseases. The possible reasons for this are many and diverse and are the subject of present research.

The uptake of antigens by fish following immersion has been demonstrated to be quite effective. The mechanism of uptake is not precisely understood but the gills appear to be the main route of entry, with the skin and lateral line system possibly playing a role. The failure to immunize fish against many diseases by the immersion method is, therefore, probably not mainly associated with inadequate uptake of the antigens but with the nature of the ensuing immune response (which is probably mainly integumentary and about which very little is known) or the inappropriate nature of the experimental challenge.

CRITERIA FOR ASSESSING VACCINE EFFECTIVENESS: SPECIFICITY AND DURATION OF PROTECTION

The most important aspect of the specific immune system exploited by vaccination is the establishment of immune memory which has a long duration (1 year to life-long). For a vaccine to have commercial application the duration of protection induced is of vital importance. Vaccines against ERM and vibriosis provide good protection for at least 1 year but little is known about the duration of immunity provided by other vaccines.

There are two aspects of the specific immune response which should be assessed, singly or in combination, before confidence can be placed in the effectiveness of a vaccine. These are specificity and duration of the protection. Unfortunately, many experimental vaccines have not been tested by these criteria as yet and a critical view should be exercised in extrapolating data which show high protection levels based on challenge with the homologous pathogen only, 4–8 weeks after vaccination. It is quite possible that a 'vaccine' may stimulate non-specific defence mechanisms which provide a degree of non-specific protection over a short time but these mechanisms lack a memory component. Indeed, many of the experimental trials with furunculosis 'vaccines' fall into this category and in view of the findings that Freund's adjuvant alone can produce at least as good protection in the short term as *A. salmonicida* antigens it would seem probable, in many cases, that the 'protection' reported is based upon non-specific mechanisms (see Chapter 8).

Hence, it is important to assess either the duration of protection over a long period or to assess the specificity of the protection (i.e. absence of protection to an unrelated pathogen) over a short period. If protection is specific it is likely to be based upon specific immune mechanisms which by implication will have a memory facet and grounds for believing the protection will have a long duration can be regarded with some confidence.

METHODS OF ASSESSING VACCINES

There are two basic factors relating to vaccine assessment: the challenge route and the method of calculating effectiveness.

The challenge route

The method of challenging fish may be important in assessing the efficacy of a vaccine. While immersion vaccination against vibriosis provides good protection to experimental water borne challenge and has proved effective against natural challenge in the field, it is much less effective when virulent-bacteria are injected into the fish. However, injection vaccination is effective against injected challenge. Injection vaccination results in high levels of circulating antibodies but immersion vaccination does not. This suggests that immersion vaccination stimulates integumentary immunity which is quite sufficient to protect against natural exposure.

Thus, it is important to assess vaccine potency by a challenge system which resembles natural exposure. Water borne challenge is therefore more appropriate than injection. Unfortunately, it is frequently very difficult to obtain sufficiently high levels of morbidity in experimental water borne challenges with many fish pathogens and researchers are often compelled to use injection challenge in assessing vaccine efficacy.

The method of calculating effectiveness

There are two accepted methods of calculating the effectiveness of a vaccine:

(a) *Relative Per cent Survival (RPS)*

This method is suitable for experimental or field trials and takes the form of challenging populations of vaccinated and non-vaccinated fish and monitoring the mortalities over a set period. The method has certain baseline criteria: at least 25 fish should be used in duplicate for each test population; the challenge should cause at least 60 per cent morbidity or mortality in control fish on a time scale similar to that of the natural disease; the cause of mortality should be determined and non-specific infections must not exceed 10 per cent in any group and the mortality rate of vaccinated fish must be below 24 per cent if the test is deemed to be positive. The calculation is made using the following formula;

$$RPS = 1 - \frac{\text{per cent vaccinate mortality}}{\text{per cent control mortality}} \times 100$$

As can be seen from the above criteria it is extremely difficult to perform these tests successfully in the field because the challenge may not be the same for all groups, the minimum mortality rate may not be achieved in controls, and intercurrent diseases may account for more than 10 per cent of the mortalities.

Even under laboratory conditions it is difficult to achieve the required level of challenge and many statistical considerations must frequently be considered in interpreting results (see Amend[8] for critical assessment of these).

(b) *Increase in Lethal Dose 50 per cent (LD_{50}).*

This test can only be conducted in the laboratory and is based on the determination of the number of virulent organisms required to kill 50 per cent of vaccinated and non-vaccinated fish. The test gives a more quantitative assessment of the vaccine potency than the RPS but is more complicated to perform. Groups of vaccinated and non-vaccinated fish comprising at least six individuals are challenged with graded doses of a virulent organism. For bacteria e.g. *A. salmonicida*, injection challenge may use doses ranging from 10 cells/fish to 10^6 cells/fish with 10-fold stepwise increases. Immersion challenge may use the same number of bacteria/ml water. Following a period relating to the natural course of the disease (for furunculosis this would be 14 days) the mortalities for each group of fish are recorded. The cause of death is determined by re-isolation of the pathogen which must be specific. The number of organisms required to kill 50 per cent (the LD_{50}) of vaccinated and non-vaccinated fish is then calculated by the formula of Reed and Muench.[9]

At least a 100-fold increase in the LD_{50} for vaccinated fish compared with non-vaccinated fish would indicate good potency of the vaccine.

A comparison of the two methods of assessing vaccine potency can sometimes reveal that an increase in LD_{50} of 10-fold in vaccinated fish compared with non-vaccinates would produce an RPS value of over 90 per cent. The latter value would be regarded as highly satisfactory but the former would cast doubt on the vaccine having a useful potency. Thus RPS values may appear high with only marginal increase in resistance to disease.

VACCINES IN THE FUTURE

Finally, mention should be made of the recent advances in immunology and biotechnology which have revolutionized the development of vaccines for use in humans. Many of the techniques used are highly sophisticated and in view of the cost of research and the relatively limited market for vaccines in fish farming their applicability in this field will be slight. Nevertheless, they should be borne in mind and a short review of some of these recent developments follows.

Genetic recombination is used to mass produce protective antigens by transferring the DNA coding the protective antigen into a suitable micro-organism which is easily cultured and which then synthesizes the antigen.

Genetic attenuation. Besides inserting pieces of DNA into the genome of a cell it is also possible to delete pieces. Specific deletion of a gene(s) which codes for a virulence factor in a pathogen, while preserving other protective antigens, allows the production of attenuated forms where reversion to the virulent wild type is impossible. Thus, the technique is valuable in production of live vaccines.

Protein engineering. The aim here is to improve the immunogenicity of an antigen or to mass produce it cheaply. Protective antigens may be made more immunogenic by cleavage of epitopes which stimulate the T suppressor cells. The protective epitopes of an antigen may be composed of a short sequence of amino acids which can be artificially synthesized quite cheaply for use as vaccines (known as 'peptide vaccines') in combination with appropriate carrier antigens.

Anti-idiotype vaccines. These are antibodies, the antigen-binding site of which mimics the structure of a protective antigen. They are produced by first raising antibodies to the appropriate antigen in a suitable animal. The specific antibodies produced (idiotypes) are then used to raise further antibodies which react with the antigen-binding site of the first (anti-idiotypic antibodies). The antigen-binding site of the anti-idiotype can be used as a vaccine since it resembles stoichiometrically the original antigen and will induce formation of antibodies which cross-react with it. Such vaccines have advantages when the protective antigen is difficult to obtain in large amounts or possesses undesirable side effects.

REFERENCES

1 Oki, A. and Sercarz, E. (1985). T-cell tolerance studied at the level of antigenic determinants. *J. Exp. Med.*, **161**, 897–911.
2 Benenson, A. S. (1970). ed. *Control of Communicable Diseases in Man*. Washington: The American Public Health Association.

3 Duff, D. C. B. (1942). The oral immunization of trout against *Bacterium salmonicida*. *J. Immunol.*, **44**, 87–94.
4 Rombout, J. H. W. M., Lamers, C. M. J., Heffrich, M. H., Dekker, A. and Taverne-Thiele, J. J. (1985). Uptake and transport of intact macromolecules in the intestinal epithelium of carp (*Cyprinus carpio*) and the possible immunological implications. *Cell & Tissue Research*, **239**, 519–530.
5 Amend, D. F. and Fender, D. C. (1976). Uptake of bovine serum albumin by rainbow trout from hyperosmotic solutions: A model for vaccinating fish. *Science*, **192**, 793–794.
6 Antipa, R. and Amend, D. F. (1977). Immunization of Pacific salmon: Comparison of intraperitoneal injection and hyperosmotic infiltration of *Vibrio anguillarum* and *Aeromonas salmonicida* bacterins. *J. Fish. Res. Board Can.*, **34**, 203–208.
7 Antipa, R., Gould, R. and Amend, D. F. (1980). *Vibrio anguillarum* vaccination of sockeye salmon, *Oncorhynchus nerka* (Walbaum) by direct and hyperosmotic immersion. *J. Fish Dis.*, **3**, 161–165.
8 Amend, D. F. (1981). Potency testing of fish vaccines. *Dev. Biol. Standard.*, **49**, 447–454.
9 Reed, L. J. and Muench, H. (1938). A simple method for estimating 50% end points. *Am. J. Hygiene*, **27**, 493–497.

2

Ontogeny of the Immune System in Teleost Fish

A. E. Ellis

INTRODUCTION

In fish, at the time of hatching, the lymphoid system is still developing and not all of the structures or functions present in the adult are present in the fry. In fact, the specific immune system is not fully mature for several weeks after the young fish has hatched. As many diseases affect fry it is important to determine the earliest age when fish can be successfully vaccinated. It is especially important because if fry are exposed to certain antigens too early, instead of priming the immune system to respond by producing antibodies, negative immunity is induced and antibody production is blocked to subsequent challenge with that antigen. This 'immune tolerance' may persist for several months and thus has serious consequences for vaccinating young fry. With other types of antigen, immune tolerance may not be induced but early exposure may have no effect on the immune response and thus vaccination with such antigens would be economically wasteful. Neverthe-less, it is in the fish farmer's interest to vaccinate fish as early as possible, not only to provide early protection but also, for immersion vaccination, the logistics of handling large numbers of small fish are simpler and more fish can be vaccinated by a given amount of vaccine.

It is likely that the earliest age a fish can be effectively vaccinated will differ between species and between vaccines and it will be important to conduct experimental trials in the future to determine the earliest age for each species of fish and for each vaccine. Presently little information is available except for salmonid fishes and carp. In this chapter the ontogeny of the defence mechanisms in fish will be described in order to understand how the nature of the developing immune system affects vaccination strategies of young fish.

MATERNAL EFFECTS

The embryo and young fry of some species receive certain humoral sub-stances which may play a defensive role and which have a maternal origin. A

FISH VACCINATION
ISBN 0-12-237485-1

number of non-specific defence factors have been identified in fish ova, including C-reactive protein-like precipitins and lectin-like agglutinins.[1] Immunoglobulin (Ig) has been found in a number of fish ova including carp,[2] and plaice.[3] Immunoglobulin does not appear to be present in the ova of salmonids. Thus, in some species at least, there is a possibility that the transfer of maternal Ig to the ova and young would result in passive immunity in the offspring which could be exploited by vaccinating gravid females.

ONTOGENY OF LYMPHOID ORGANS

The major lymphoid organs in teleost fish are the thymus, kidney and spleen. The thymus is a paired, bilateral organ situated beneath the pharyngeal epithelium dorso-laterally in the gill chambers. It is composed mainly of developing lymphocytes. It is regarded, as in other vertebrates, as a primary lymphoid organ where the pool of virgin lymphocytes is produced and which then emigrates to join the peripheral pool of lymphocytes in the circulation and other lymphoid organs. The thymus appears to have no executive function i.e. it does not participate in antibody production or uptake of antigens.

The kidney is the main antibody-producing organ. It contains a generalized haemopoietic tissue rich in lymphocytes and plasma cells (antibody-producing cells). It is also a filtration organ containing many macrophages which phagocytose antigens.

The spleen contains fewer haemopoietic and lymphoid cells than the kidney, being composed mainly of blood held in sinuses. However, it contains specialized capillary walls, termed ellipsoids, which are composed of reticulin fibres and macrophages. The latter are highly phagocytic while the reticulin fibre network is specialized for trapping immune-complexes (complexes of antibody and antigen).[4] The function of this in fish is unclear, but in mammals an analogous process of antigen trapping is concerned with the development of immune memory.

The thymus is the first lymphoid organ to develop lymphocytes. It arises within the epithelium of the branchial cavity[5] (Fig. 2.1) and in young salmonids the fully differentiated thymus is separated from the external environment only by a single layer of simple epithelial cells which in rainbow trout possesses pores up to 20 μm in diameter. In older fish these pores close and the epithelium thickens. In carp, the thymus is initially superficial but quickly becomes embedded in deeper tissues.

The time of appearance of lymphocytes in the blood circulation and lymphoid organs is compared for the Atlantic salmon, rainbow trout and carp in Table 2.1. Following the differentiation of large numbers of cells in the thymus, lymphocytes then appear in the blood and the kidney at the same time. The kidney then rapidly becomes rich in lymphoid cells. The spleen develops later. The exact timing of the lymphocyte differentiation varies between species and this is probably related to the rate of growth and general development. What is apparent is that morphologically mature small lymphocytes appear just before or just after hatching.

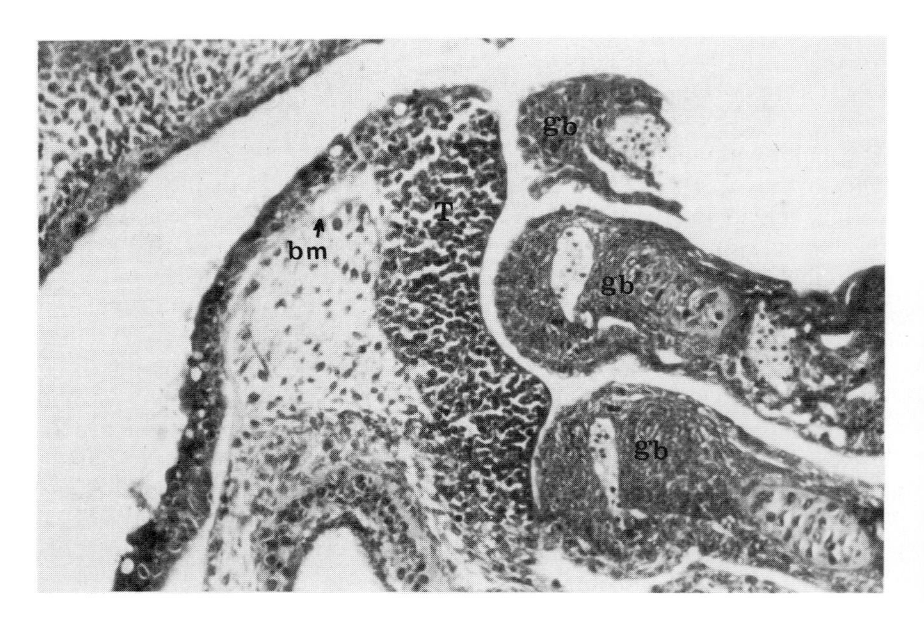

Fig. 2.1 The thymus (T) of Atlantic salmon embryo 8 days pre-hatch. The organ develops within the pharyngeal epithelium, external to the basement membrane (bm) and is covered externally by a single layer of squamous epithelium. gb, primary gill bars.

Table 2.1. *Appearance of lymphocytes in the developing lymphoid organs of three teleost species*

	Salmon[5] 4–7°C	Rainbow trout[16,22] 14°C	Carp[15] 22°C
Thymus	22 days pre-hatch	3–5 days post-hatch	5 days post-hatch
Blood	14 days pre-hatch	5–6 days post-hatch	7–8 days post-hatch
Kidney	14 days pre-hatch	5–6 days post-hatch	7–8 days post-hatch
Spleen	42 days post-hatch	21 days post-hatch	8–9 days post-hatch

During the first few weeks post-hatching the rate of growth of the lymphoid tissues in rainbow trout is faster than the rest of the body and the weight of the lymphoid organs, relative to body weight, reaches a peak at 2 months of age, when fish are 0.5 g.[6] Thereafter, while the lymphoid organs continue to grow, they do so more slowly than the rest of the body so that their relative weights decrease with age (Fig. 2.2). During the first few months there is intense mitotic division of lymphocytes in the thymus and then a decrease. It is believed that many thymocytes emigrate to peripheral organs during the first 2–3 months post-hatching and signs of involution of the thymus appear at about 9 months.

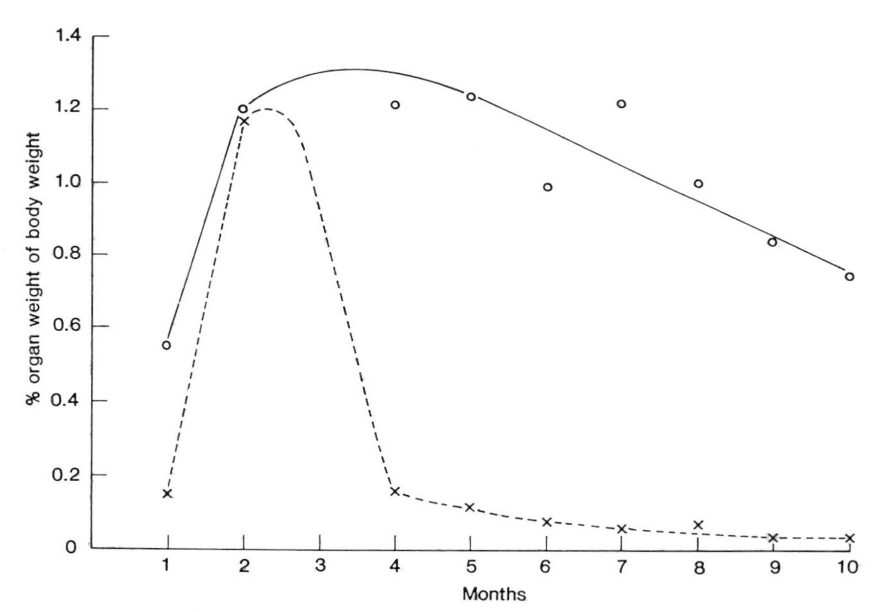

Fig. 2.2 Weight of thymus and kidney as a per cent of body weight in rainbow trout of ages 1–10 months. (Data from Ref. 6). ×, thymus; ○, kidney.

The development of the lymphoid organs correlates better with the weight of the fish than with age and therefore appears to be a function of the rate of growth of the fish. As will be seen later, this has significance for vaccinating fry as young as possible.

ONTOGENY OF THE PHAGOCYTIC SYSTEM

The development of this system has been studied in rainbow trout injected intra-peritoneally with suspensions of carbon particles.[7] In the adult trout the major phagocytic sites are in the kidney, spleen (ellipsoids) and the epicardium. At 4 days post-hatch the spleen is not yet present and the kidney still lacks lymphoid cells. Carbon particles are phagocytosed by macrophages in the gills and connective tissues of the skin and gut. Only a few phagocytes are present in the kidney. At 18 days post-hatch the kidney is fully lymphoid and a rudimentary spleen is present. Carbon particles are predominantly phagocytosed by macrophages in the gills and the kidney but small amounts are phagocytosed in the spleen and epicardium. In the next few weeks, the numbers of gill macrophages decline and the uptake of carbon particles takes on the pattern seen in the adult.

These findings suggest the possibility that during the stages when the lymphoid organs are developing, the entry of foreign material from the environment may be prevented from entering the systemic circulation of the young fry by macrophage populations within the integument. Once the peripheral organs are fully lymphoid the gill macrophages are replaced by lymphoid organ macrophages. These developmental stages may have significance in the development of immunological responsiveness (see below).

ONTOGENY OF LYMPHOCYTE MATURATION

Although morphologically identifiable small lymphocytes are present before or very soon after hatching it is apparent that they are not yet functionally mature. In the adult, peripheral lymphocytes carry on their surface certain antigens related to the cell's function. The presence of these antigens can be used as a marker for differentiating functionally immature from mature cells and, in mammals, for distinguishing between populations of lymphocytes with different functions. In adult fish, all lymphocytes carry an antigen marker which cross reacts with certain epitopes on serum antibodies. This marker is termed surface immunoglobulin (sIg) and is associated with the antigen recognition site of the lymphocyte. Appearance of this marker on lymphocytes during ontogeny has been studied. In Atlantic salmon, although the thymus and kidney are fully lymphoid at the time of hatching, prior to day 35 post-hatch the lymphocytes lack sIg. By day 48 post-hatch the majority of lymphocytes possess the marker.[5] This coincides with first feeding and it would seem highly unlikely that salmon fry are capable of producing an antibody response prior to this time.

This technique for studying the early maturation of lymphocytes has been refined by the use of monoclonal antibodies which have exquisite specificity for certain markers. On the basis of their reactivity with carp serum Ig and carp thymocytes (T), three monoclonals were denoted $Ig^+ T^+$, $Ig^- T^+$ or

$Ig^+ T^-$.[8] The development on lymphoid cells in the thymus and kidney of carp fry of markers which react with these monoclonal antibodies is shown in Table 2.2.

Table 2.2 *The appearance of surface markers on lymphoid cells during the development of the lymphoid organs in carp at 21°C (data from Ref. 8)*

Days post-hatch	Thymus		Kidney		
	$I_g^+ T^+$	$I_g^- T^+$	$I_g^+ T^+$	$I_g^- T^+$	$I_g^+ T^-$
2	+	−	−	−	−
5	+	+	+	−	−
10	+	+	+	+	−
14	+	+	+	+	+
19	+	+	+	+	+

It is apparent that the different determinants appear on lymphoid cells at different times during ontogeny. The T^+ marker appears first in the thymus and later in the kidney. The $Ig^+ T^-$ cells (possibly B cells), which are absent from the thymus, appear quite late in the kidney just preceding cells which stain for this determinant within the cytoplasm and are therefore, presumably, plasma cells.

Thus, although lymphocytes are present early in development their functional maturation takes a further period of time before they are capable of executing an immune response.

ONTOGENY OF IMMUNE RESPONSIVENESS

Humoral immunity and tolerance

There are two important factors which influence the character of the antibody response in young fish: the nature of the antigen and the route of exposure (Table 2.3). Antibody production in fry to antigens which in mammals are thymus (T)-independent (e.g. bacterial lipopolysaccharide) precedes thymus (T)-dependent antigens (e.g. most soluble protein antigens). Rainbow trout fry (at 4–6°C) do not produce an antibody response when immunized by direct immersion with *Aeromonas salmonicida* bacterin at 1 week post-hatch, but they do when 4 weeks of age (0.13 g). However, evidence for memory development (as measured by an increased secondary response) was not seen until fish were immunized at 8 weeks of age (0.26 g). The ability to respond to a T-dependent antigen, human gamma globulin

Table 2.3 *Antibody production and establishment of memory or tolerance in rainbow trout and carp of different ages immunized with a thymus-dependent (HGG) or independent* (A. salmonicida) *antigen by two different routes*

Age (weeks) when first immunized	Rainbow trout (immersion 6°C injection 14°C)				Carp (20°C)			
	A. salmonicida		HGG		*A. salmonicida*		HGG	
	Immersion	Injection	Immersion	Injection	Immersion	Injection	Immersion	Injection
1	–	–	–	–	–			
4	$+(M^-)$	$+(M^-)$	–	–(T)	+	$+(M^+)$	$-(M^-)$	–(T)
8	$+(M^+)$	$+(M^+)$	$+(M^+)$	$+(M^+)$	$+(M^+)$		$-(M^-)$	$+(M^+)$

Key: – no antibody produced; + antibody produced; M^- no memory; M^+ memory established; T tolerance. (Data from Refs. 9, 10, 11, 13)

(HGG), was somewhat later with the production of antibody and the establishment of memory first developing when the fry were immersion immunized at 8 weeks of age.[9] When these antigens were injected into young trout, similar results were obtained with the bacterial antigen but tolerance to HGG was induced when injected into fry 21 days post-hatch. This immune tolerance persisted for at least 8 weeks. However, fish responded positively when injected at 8 weeks of age.[10] These results imply that B cells and T suppressor cells may become functionally mature at about 4 weeks of age but T helper cell function and memory cells mature later at 8 weeks.

Studies on carp fry have similarly indicated that antibody production to T-independent antigens precedes T-dependent ones and early exposure to T-dependent antigen can result in the induction of immune tolerance. When 4 week old carp were injected with *Aeromonas salmonicida* bacteria a good antibody response was achieved as well as the establishment of memory.[11] However, upon injection of HGG in 4 week old carp, no antibody was produced and a state of antigen specific tolerance was established which persisted for at least 23 weeks. When injected at 9–10 weeks of age the fry developed a normal antibody response and memory.[12,13]

A state of tolerance was not established when fry were immunized by immersion in antigen even as early as 1 week post-hatch. In fact, immersion in HGG once a week for weeks 4–8 post-hatch inclusive had no effect i.e. no antibody was produced and no memory or tolerance.[13] On the other hand, immersion immunization in *A. salmonicida* bacteria under the same regime beginning at 4 weeks of age, resulted in antibody production and the establishment of memory.[11]

It is not entirely clear why injection of soluble HGG in carp and trout of 4 weeks of age induces tolerance whereas immersion in this antigen does not. A possible explanation is that soluble antigens are rather poorly taken up by the gills of young fry and/or most of the antigen which does enter the gills is prevented from reaching the lymphoid tissues by being phagocytosed by the macrophages in the gills (see ontogeny of phagocytic system above). Soluble antigens are not as readily taken up as particulate antigens and attachment to particulate carriers greatly facilitates uptake.[14] Thus when carp fry were immersion immunized in HGG linked to latex particles, although a state of tolerance was not produced, it did result in a slight lowering of the antibody response to a subsequent injection of HGG compared with control fry.[11] It seems possible, therefore, that the more efficient the uptake of HGG by young carp fry, the more likely it is to induce tolerance. Nevertheless, this begs the question as to why HGG administered by immersion should be poorly taken up by young fry when it is taken up by 8 week old fry and results in a positive immune response. A possible explanation is that the large numbers of macrophages present in the gills of young fry destroy the antigenicity of HGG entering through the gills, so that the antigen does not gain access to the yet developing lymphoid tissues in the kidney.[7] Once the lymphoid tissues are functionally mature (by 8 weeks of age in carp and trout) the numbers of macrophages in the gills have markedly declined so that antigens entering via the gills can gain access to the kidney tissues through the circulation. Thus it is possible that the gill macrophages of young fry provide a means of preventing the access of certain types of antigen to the developing lymphoid tissues which otherwise would induce immune tolerance. But how could T-independent antigens be handled differently in this respect?

Clearly it is important to establish if the effects described above are generally valid for a wide range of thymus-dependent and independent-type antigens. The implications are clear; vaccination with T-independent antigens can be performed in rainbow trout and carp as early as 4 weeks post-hatch, but vaccination with T-dependent antigens may need to be postponed for at least 8 weeks to avoid the danger of inducing immune tolerance instead of desired immunity (Table 2.3).

CMI

Development of the CMI response in young fish has been investigated using two tests: the mixed leucocyte response (MLR) and the allograft rejection response. In the Atlantic salmon the MLR develops at day 42 post-hatch, coincidentally with first feeding.[5] Skin rejection responses in carp and rainbow trout fry are developed by 16 days[15] and 14 days post-hatch[16] respectively. In the trout, grafting at 5 days post-hatch resulted in incomplete rejection by day 30 post-grafting but the experiment was not continued to establish if allograft application at this age eventually resulted in rejection or tolerance.

The ontogeny of cell-mediated immune memory has been investigated following the application of first set skin grafts to 26 days old rainbow trout and 16 days old carp.[17] In both cases memory developed. It is thus apparent that in trout and carp the CMI system has fully matured by 2–4 weeks post-hatch and suggests that this system, with the production of cytotoxic T-like cells, matures a little earlier than the humoral immune response particularly to the T-dependent antigen, HGG.

THE EARLIEST TIME TO VACCINATE

Extensive data on the earliest time to vaccinate is only available for salmonids immunized against enteric redmouth and vibriosis. There are three aspects of practical importance: the earliest age at which fry develop maximum protection, the duration of that protection, and the effect of temperature.

At 10°C most species of salmonids do not develop protective immunity when vaccinated at or below 0.5 g. Thereafter, increasing levels of protection develop so that maximum levels of protection are achieved in most species at 1 g and in all species tested by the time they are 2.5 g.[18] It is the size of the fish that appears important. When fish are reared at different temperatures to vary their growth rate, protective immunity was shown to correlate with body size, not age.

Similar work on rainbow trout vaccinated against vibriosis[19] basically confirmed these findings though the minimum size to achieve acceptable levels of protection was reported as 0.5 g (10 weeks post-hatch) in fish which were held at 10°C and which began feeding at 4 weeks post-hatch. This correlated with the maximum relative size of the lymphoid organs and it has been suggested that the development of protective immunity is dependent upon the presence of a critical number of lymphocytes.

Duration of protection against ERM and vibriosis increases with age (size) when fish are vaccinated. Typical figures are 120 days for 1 g fish; 180 days for 2 g fish and over 1 year for 4 g fish, which is the duration existing in adults.[20]

Temperature has a marked effect on the immune response in fish so that the higher the temperature, within the physiological range, the higher the level of the immune response (see Ch. 3). Physiologically low temperatures retard the rate of antibody production and may inhibit the establishment of immune memory. Unfortunately, no studies have been done to assess the efficacy of vaccinating fry of different sizes at different temperatures. Obviously, temperature affects the growth rate of fry and low temperatures will prolong the period required for the fish to achieve the critical developmental stage (size) at which it becomes immunologically competent. The lowest temperature at which the effectiveness of vaccination has been critically studied is 6°C in grower rainbow trout (100–200 g in weight) vaccinated against vibriosis.[21] The level of protection achieved was equal to that of fish held at 10°C.[19] Thus, for vaccination against vibriosis, ambient temperature is not expected to be a factor which greatly restricts the vaccination of fish provided they are over the critical size.

SUMMARY AND CONCLUSIONS

Studies on the development of the immune system in trout and carp have shown that cytotoxic cells and antibody production to T-independent antigens develop at about 4 weeks post-hatch. The ability to establish immune memory and to produce antibody to T-dependent antigens develops by about 8 weeks post-hatch. Exposure by injection (though not immersion) to T-dependent antigen before fry are 8 weeks old leads to immunological tolerance.

Protective immunity in salmonids vaccinated against vibriosis or ERM develops at about 8 weeks post-hatch but does not reach economically acceptable levels until fry are 0.5–1.0 g (about 10 weeks post-hatch for fish kept at 10°C). The size rather than the age of the fry is the critical factor. The ability of salmonid fish to develop prolonged duration of protection (long-term memory) upon vaccination is fully mature when fry are about 4 g.

REFERENCES

1 Ingram, G. A. (1980). Substances involved in the natural resistance of fish to infection – A review. *J. Fish Biol.*, **16**, 23–60.

2 Van Loon, J. J. A., van Oosterom, R. and van Muiswinkel, W. B. (1981). Development of the immune system in carp. In: *Aspects of Developmental and Comparative Immunology,* vol 1, ed. J. B. Solomon, Oxford: Pergamon Press, pp. 469–470.

3 Bly, J. E., Grimm, A. S. and Morris, I. G. (1986). Transfer of passive immunity from mother to young in a teleost fish: haemagglutinating activity in the serum and eggs of plaice. *Pleuronectes platessa* L. *Comp. Biochem. Physiol.,* **84A**, 309–314.

4 Ellis, A. E. (1980). Antigen trapping in the spleen and kidney of the plaice, *Pleuronectes platessa. J. Fish Dis.,* **3**, 413–426.

5 Ellis, A. E. (1977). Ontogeny of the immune response in *Salmo salar*. Histogenesis of the lymphoid organs and appearance of membrane immunoglobulin and mixed leucocyte reactivity. In: *Developmental Immunobiology,* ed. J. B. Solomon and J. D. Horton, Elsevier/North Holland, Biomedical Press, pp. 225–231.

6 Tatner, M. F. and Manning, M. J. (1983). Growth of the lymphoid organs in rainbow trout, *Salmo gairdneri*, from one to 15 months of age. *J. Zool., Lond.,* **199**, 503–520.

7 Tatner, M. F. and Manning, M. J. (1985). The ontogenetic development of the reticuloendothelial system in the rainbow trout, *Salmo gairdneri* Richardson. *J. Fish Dis.,* **8**, 35–41.

8 Secombes, C. J., Van Groningen, J. J. M., Van Muiswinkel, W. B. and Egberts, E. (1983). Ontogeny of the immune system in carp (*Cyprinus carpio* L.). The appearance of antigenic determinants on lymphoid cells detected by mouse anti-carp thymocyte monoclonal antibodies. *Dev. Comp. Immunol.,* **7**, 455–464.

9 Tatner, M. F. (1986). The ontogeny of humoral immunity in rainbow trout, *Salmo gairdneri. Vet. Immunol. Immunopathol.,* **12**, 93–105.

10 Manning, M. J., Grace, M. F. and Secombes, C. J. (1982). Ontogenetic aspects of tolerance and immunity in carp and rainbow trout: Studies on the role of the thymus. *Dev. Comp. Immunol. Suppl.,* **2**, 75–82.

11 Manning, M. J. and Mughal, M. S. (1985). Factors affecting the immune responses of immature fish. In: *Fish and Shellfish Pathology,* ed. A. E. Ellis, London: Academic Press, pp. 27–40.
12 Mughal, M. S. (1984). Immunological memory in young fish and amphibians: Studies on *Cyprinus carpio, Chelon lasosus* and *Xenopus laevis*. PhD. Thesis, Plymouth, UK.
13. Mughal, M. S. and Manning, M. J. (1985). Antibody responses of young carp (*Cyprinus carpio*) and grey mullet (*Chelon labrosus*) immunized with soluble antigen by various routes. In: *Fish Immunology,* ed. M. J. Manning and M. F. Tatner, London: Academic Press, pp. 313–325.
14 Smith, P. D. (1982). Analysis of the hyperosmotic and bath methods for fish vaccination – comparison of uptake of particulate and non-particulate antigens. *Dev. Comp. Immunol. Suppl.,* **2**, 181–186.
15 Botham, J. W. and Manning, M. J. (1981). Histogenesis of the lymphoid organs in the carp, *Cyprinus carpio* L. and the ontogenic development of allograft reactivity. *J. Fish Biol.,* **19**, 403–414.
16 Tatner, M. F. and Manning, M. J. (1983). The ontogeny of cellular immunity in the rainbow trout, *Salmo gairdneri* Richardson, in relation to the stage of development of the lymphoid organs. *Dev. Comp. Immunol.,* **7**, 69–75.
17 Botham, J. W., Grace, M. F. and Manning, M. J. (1980). Ontogeny of first set and second set alloimmune reactivity in fishes. In: *Phylogeny of Immunological Memory,* ed. M. J. Manning, Elsevier/North Holland, Biomedical Press. pp. 83–92.
18 Johnson, K. A., Flynn, J. K. and Amend, D. F. (1982). Onset of immunity in salmonid fry vaccinated by direct immersion in *Vibrio anguillarum* and *Yersinia ruckeri* bacterins. *J. Fish Dis.,* **5**, 197–205.
19 Tatner, M. F. and Horne, M. F. (1983). Susceptibility and immunity to *Vibrio anguillarum* in post-hatching rainbow trout fry, *Salmo gairdneri* Richardson 1836. *Dev. Comp. Immunol.,* **7**, 465–472.
20 Johnson, K. A., Flynn, J. K. and Amend, D. F. (1982). Duration of immunity in salmonids vaccinated by direct immersion with *Yersinia ruckeri* and *Vibrio anguillarum* bacterins. *J. Fish Dis.,* **5**, 207–213.
21 Horne, M. T., Tatner, M., McDerment, S., Agius, C. and Ward, P. (1982). Vaccination of rainbow trout, *Salmo gairdneri* Richardson, at low temperatures and the long persistence of protection. *J. Fish Dis.*, **5**, 343–345.
22 Grace, M. F. and Manning, M. J. (1980). Histogenesis of the lymphoid organs in rainbow trout, *Salmo gairdneri. Dev. Comp. Immunol.,* **4**, 255–264.

3

Optimizing Factors for Fish Vaccination

A. E. Ellis

INTRODUCTION

There are a number of factors which affect the development of an immune response. Factors inherent to the fish (intrinsic factors e.g. age, health status) and those which relate to the vaccine, vaccination method or environment (extrinsic factors e.g. dose of antigen, route of administration, adjuvants, temperature), are important to consider if vaccination is to be effective. On a farm it is not always possible to provide optimal conditions for attaining maximum efficacy of a vaccine but adequate protection against disease may nevertheless be achieved. However, if the conditions under which fish are vaccinated are severely inadequate, immunization will fail and it is important that the fish farmer is aware of the limits within which a vaccine will be effective. In the previous chapter, the importance of the age, or more accurately the size, of fish fry was described. In general, salmonids do not develop a very effective level of immunity when vaccinated below 0.5–1 g and do not develop prolonged immune memory until about 4 g. However, for certain diseases significant mortalities may occur in fish of below 4 g and even though induction of immune memory is poor in smaller fish it is still possible to vaccinate them provided they are above 1 g.

In this chapter some of the important factors which affect the development of effective immunity will be described. Some are mainly the concern of the fish farmer relating to fish husbandry and others of the fish immunologist or vaccine manufacturer, relating to the nature and presentation of the vaccine.

HUSBANDRY FACTORS

Stress

The immune system, like other physiological systems, is compromised by poor health. Stressful circumstances are known to depress certain aspects of

FISH VACCINATION
ISBN 0-12-237485-1

immune responsiveness.[1] The stress response involves physiological and behavioural reactions which may help the fish adapt to a new situation but if the stress is severe or prolonged, it may exceed the capacity of the fish to adjust resulting in a general breakdown of the immune (and other) systems. Thus, if fish are affected by an infective agent or suffering ill health from a poor quality environment (e.g. low oxygen, high ammonia from excrements, suspended solids) the efficacy of vaccination will be decreased. The suppressive effects of stress on the immune response are mediated by hormones, principally the corticosteroids. There is an increase in these hormones at certain times of the life cycle in some species of fish e.g. at smolting in Atlantic salmon and during sexual maturation of male salmonids. Sexually maturing and mature male salmonids appear to have an increased susceptibility to infectious diseases and it has been suggested they may be immunologically suppressed resulting from the high level of corticosteroids.[2] No work has been performed to assess the efficacy of vaccinating mature male fish but the indications are that success may be slight. Many aspects of fish farming which involve the handling of fish are known to cause stress (e.g. transport, grading, anaesthesia), and such treatments are known to suppress antibody production markedly.[3]

High population density is another important factor which besides having significant repercussions on water quality, may also suppress the immune response. Many fish species living in groups organize themselves into dominance hierarchies and subordinate members have persistently elevated corticosteroid levels in the blood. The poor growth rates of such individuals and in some cases an ultimate stress induced death (as in eels[4]) implies that immune responsiveness would also be markedly impaired. In carp populations only the dominant members produced antibodies to experimental infection with trypanosomes.[5]

Further to the direct effects of stress resulting from crowding some species of fish release pheromones when stressed, inducing a stress response in others. For example, a substance in the water in which carp were kept under crowded conditions caused depression of heart and growth rates of uncrowded carp (see Ref. 1). A similar pheromone is released by blue gouramis, *Trichogaster trichopterus*, under crowded conditions and has been shown to reduce the immune responsiveness to IPN virus. The immune suppression was removed if the water in which the fish were kept was extracted with methyl-chloroform (see Ref. 1). Stress pheromones are also released by tilapia (*Sarotherodon mossambica*), causing the release of pharmacologically active substances in other individuals which may result in their death.[6] Although salmonids are stressed by overcrowding, these species do not appear to produce stress pheromones.[7]

From the above, for the purposes of vaccinating fish, it is apparent that stress should be avoided. But there are many different kinds of stress responses and there are many differences between fish species as to what they find stressful. A good fish farmer should know when the fish are contented or not and current scientific wisdom would advise that if the fish are suffering a fairly severe stress for any reason (immediately after traumatic transport; severe spates; attacks by predators; an infectious disease etc.) it would not be a good time to vaccinate. Many studies have shown that following transient stress fish return to a normal physiological state in about 10 days, as judged by certain biochemical and endocrine criteria.[7] It would, therefore, appear prudent for a fish farmer to acclimate fish after stocking facilities etc. for this period of time before vaccinating them.

Finally, on this note of stress, with the selective breeding of fish to obtain improved performance under aquaculture conditions, the process of domestication will result in the fish becoming less stressed by standard husbandry practices.

Diet

As mentioned above, efficient immune responses to vaccination depend upon the fish being healthy and in good condition. If protective factors are to be synthesized and cell function maintained, the fish must be provided with balanced diets to meet nutritional requirements. In the main this has been achieved for carp, rainbow trout, Atlantic salmon and channel catfish and a few other species. However, the nutritional requirements for all nutrients are obtained from dose response curves in which the main response measured has been weight gain although some attention has been paid to morbidity and pathology. Very few studies have, however, been concerned with the effect of nutrients on the immune response. The only studies so far relate to levels of ascorbic acid (Vit. C) and α-tocopherol (Vit. E).

Vitamin C

Levels of vitamin C greater than the maximum dietary requirement for normal growth increased the resistance of channel catfish fingerlings to *Edwardsiella tarda* and *E. ictaluri* infection especially at low (23°C) temperatures.[8] The antibody response to *E. ictaluri* antigens was lower in channel catfish given Vit. C deficient diets than in those given normal doses (30–300 mg Vit. C/kg diet). Megadose amounts (3000 mg Vit. C/kg diet) significantly enhanced antibody responses still further. Vit. C deficiency also reduced the phagocytic activity of blood phagocytes towards *E. ictaluri*, though megadose amounts of Vit. C did not enhance phagocytic activities compared with normal dose regimes. Vit. C levels did not appear to affect the bactericidal activity of the phagocytes.[9]

These results indicate that normally recommended Vit. C in diets for channel catfish may be a limiting factor for optimal functioning of the immune response and other non-specific defence mechanisms particularly at low temperatures when supplementation of the diet has been recommended.

Vitamin E

Rainbow trout fed a Vit. E deficient diet for 12–17 weeks had reduced antibody responses and phagocytic activity compared with control fish fed a diet containing 40 Iu Vit. E/100 g diet.[10] It is of interest, in these experiments, that the fish fed Vit. E deficient diets appeared healthy over this entire period. Indeed, rainbow trout fed a Vit. E deficient but otherwise normal diet, do not differ in growth rate from control fish and certain biochemical indices (e.g. erythrocyte fragility) are needed to define Vit. E requirements. Using such techniques, rainbow trout have been shown to have a higher requirement for polyunsaturated fats and Vit. E during winter months, i.e. low temperatures.[11]

These results indicate that the immune response may be very sensitive to Vit. E levels even though otherwise the fish may appear healthy. It is possible that, particularly at low temperatures, the levels of Vit. E in commercial diets (which normally range from 7.5–40 Iu/kg diet) may be inadequate for optimal immune responsiveness.

Further work is required to confirm these results and to extend studies on the number of dietary factors which may affect immune responsiveness. Besides Vit. C and E, other factors known to affect the immune response in mammals include Vit. A, zinc and the type of dietary fats. The outlook at the moment appears to be that such studies will lead to improvements in fish diets leading to improved disease resistance in vaccinated as well as non-vaccinated fish.

Pollutants

Sublethal doses of toxic agents can have effects upon the defence mechanisms in fish that may ultimately be as harmful as direct toxic doses by increasing susceptibility to disease. Various pollutants are known to suppress the immune response of fish and/or to increase disease susceptibility (Table 3.1). The mechanism of action is little understood but there are three possible causes:

(a) the pollutant may inhibit uptake of the antigen from bath immunizations by affecting gill permeability (e.g. phenol[12]).
(b) induction of persistent stress in the fish causing suppression of the immune response.

(c) directly interfering with cells or their products involved in defence (e.g. many of the heavy metals[13]).

Table 3.1 *Effect of various pollutants on increased susceptibility to disease and decreased immune responsiveness in fish (Data from Refs. 12, 13, 46.)*

Pollutant	Concentration (ppm)	Fish	Increased susceptibility to:	Decreased immune response to:
Cadmium	sublethal	cunner	bacteria	CMI suppressed
Chromium		carp, brown trout		viruses
Copper	0.01–0.03	Coho salmon		*Vibrio*
	0.03–0.06	eel		vibriosis
	0.29	carp, brown trout		virus
Zinc	0.3	Zebrafish		bacteria
Methylmercury	9	blue gourami		bacteria and viruses
Phenol	12.5	carp		*Aeromonas*
	10	rainbow trout		*Yersinia ruckeri*
Pesticides:				
Chlorinated hydrocarbons	sublethal	various	various diseases	
PCB	0.0001–0.01	various	various diseases	

Antibiotics

Antibiotics are known to affect the immune response in mammals but the effects are variable depending on the antibiotic, the aspect of defence and the species. The effect may be enhancement, suppression or none. Little is known about this subject in fish, though concern has been expressed about the immune suppressive effect of oxytetracycline (OTC) in carp and rainbow trout. However, these studies have involved either the treatment of fish with much higher doses than would occur during therapeutic treatment of diseased fish on fish farms or have made inadequate studies of the immune response. Nevertheless, the present information is sufficient to warrant further investigation as it may have importance for vaccinating fish which are being treated with OTC.

When carp were injected intramuscularly with rabbit erythrocytes a high primary humoral response was detected on day 9. However, if fish were treated with OTC either orally (50 mg/kg fish/day throughout the experiment)* or by injection (60 mg/kg fish every 3rd day) beginning 15 days

* Recommended therapeutic doses for treating fish with OTC are orally: 50 mg/kg fish/day for 10 days; injection: 10–20 mg/kg (usually single injection).

before injection of antigen, the humoral response was suppressed by over 90 per cent when assayed on day 9.[14] Similar results were obtained with rainbow trout where feeding the same dose of OTC for 14 days prior to injection of *Yersinia ruckeri* O antigen reduced the humoral response by 75 per cent. Feeding OTC for 14 days after immunization produced over 90 per cent reduction.[15] Unfortunately, in the case of the work on carp the OTC treatment was for a longer time than would occur in a therapeutic administration on a farm and in the case of the rainbow trout the antibody response was not apparently assessed after the expected peak time. This may be important, since in experiments with carp injected every third day over an 11 day period following immunization with sheep erythrocytes, the antibody response was delayed for 2–4 days in comparison with saline injected control fish but otherwise the titres were as high and were maintained for the same period.[16]

In a study of the effect of OTC on the development of protection against vibriosis in rainbow trout, the results appear more reassuring.[17] Here, OTC treatment (oral, 75 mg/kg fish/day for 8 days) given 26 days after dip vaccination or beginning 6 h prior to low dose infection of vaccinated or non-vaccinated fish did not interfere with the development of protection to a subsequent high dose bath challenge 30 days later.

Obviously, the effects of OTC on the immune response in fish requires further investigation. The only other antibiotic for which relevant data are published is the potentiated sulphonamide R05–0037 (fed at 5 mg/kg fish/day for 5 days) which had no suppressive effect on the antibody response of rainbow trout to *Y. ruckeri* O antigen.[12]

Environment

Besides the factors relating to a suboptimal environment for health which have been discussed above under the heading of Stress, there are two aspects of a healthy environment which can affect the immune response and which may be relevant to the fish farmer deciding when to vaccinate fish. These are temperature and season.

Temperature

Many studies have shown that the immune response (CMI and humoral) of fish is slower at low temperatures. In general, the best responses are obtained at the normal summer temperatures of the species concerned. These, of course, vary considerably according to whether the fish is a coldwater, temperate or warmwater species. Temperatures which are low within the normal physiological range of a particular species of fish (e.g. less than 6°C for rainbow trout or 12°C for carp) can depress the immune response in a variety of ways.[18] However, the extent to which this may occur in the field situation is unclear since most experimental work has failed to acclimatize fish to lower temperatures for an appreciable period of time and it has recently been shown that biochemical and immunological adaptation to lower temperatures does occur in fish if acclimated for at least 2–3 weeks. The mechanism of adaptation may be to increase the relative amounts of polyunsaturated fatty acids in the cell membrane phospholipids which has the adaptational effect of re-establishing cell membrane fluidity at lower temperatures.[19] This explains why the dietary requirements for polyunsaturated fatty acids increases for trout at low temperatures (see diet above).

Unfortunately, few studies have been performed on the *in vivo* immune responsiveness of fish at low physiological temperatures. The only study relevant to vaccination demonstrated comparable efficacy of immersion vaccination of rainbow trout against vibriosis when carried out on fish at 6°C[20] compared with 10°C.[21] However, it is possible such reassurance will not be available for some other vaccines. The protective antigen in *Vibrio* vaccines appears to be the lipopolysaccharide which is a T-independent antigen and low temperatures seem to preferentially depress the immune response to T-dependent antigens.[22] In carp, which were acclimated to 12, 16, 18, 20 or 24°C for at least 2 weeks prior to immunization with sheep erythrocytes, the peak of the primary antibody response was not decreased but it was delayed (day 9 at 24°C; day 49 at 12°C). Of more significance to vaccination is that at temperatures at or below 18°C the induction of immune memory was absent.[23] This would be a significant limitation to effective vaccination where protective antigens are T-dependent and may mean that vaccinating fish below a certain temperature would provide only short-term protection. However, normal secondary immune responses can be elicited at low temperatures in fish which are immunized at higher temperatures.[18] This gives the possibility to certain fish farmers of artificially raising the temperature of fish for a short period (maybe 2–3 weeks) at the time of vaccination to allow development of immune memory which can then function at a low temperature.

Seasonal effects

Many physiological systems have rhythmical activities on a daily, monthly or annual basis. Some evidence exists to indicate that the immune system of fish may have seasonally associated depressed periods which are not simply associated with low temperatures.[24] Rainbow trout, held at 18°C throughout the year, developed lower antibody titres to *Aeromonas salmonicida* when immunized prior to winter compared with fish immunized prior to spring. Similar results were obtained in the rockfish, *Sebastiscus marmoratus* immunized with sheep erythrocytes.[25] Furthermore, the reactivity of mature female rockfish was lower than that of males or immature females in the spawning season (winter). A slight alleviating effect on depression of antibody production during the winter season was achieved by extending the day length period.

It would appear that this is an important area for future research as it may be possible, by simple manipulation of the environment, to optimize conditions for vaccinating fish which are subject to seasonal variation in their immune responsiveness.

VACCINE-RELATED FACTORS

There are a number of factors relating to the nature of the vaccine or vaccination method which affect the immune response; the fact that T-dependent antigens, though not T-independent ones, can induce long-term tolerance when injected into young fish, though not when exposed by immersion, has been discussed in Chapter 2. This example shows the interdependence of many factors on the outcome of the immune response, in this case the interdependence of fish size, nature of the antigen and route of administration. This interdependence of factors is one of the major problems for fish immunologists attempting to develop vaccines because a test vaccine may fail only because one important factor has been inadequate for inducing effective protection. The factors discussed below are interdependent in one way or another.

Antigen dose

The level of antibody production in the primary immune response is (within limits) usually correlated with the dose of antigen administered. In certain cases, very high doses of antigen injected into fish can induce tolerance e.g. injection of 10 mg bovine serum albumin intracardially in carp.[26] However, for the induction of memory, the effect of antigen dose appears to depend on the nature of the antigen.

Nature of the antigen

In carp immunized by injection of sheep erythrocytes, induction of immune memory was most effective at low and medium doses, although it took longer to develop with the low dose.[23] On the other hand, the maximum memory to *Aeromonas hydrophila* injected intramuscularly into carp was directly correlated with antigen dose and low dose priming induced only weak memory formation.[27]

The interdependence of the nature of the antigen and temperature has been referred to above. Generally the antibody response to T-independent antigens is only marginally affected by low physiological temperatures whereas that to T-dependent antigens is markedly slower and without memory formation.

The form in which an antigen is administered may be very important. Generally, soluble protein antigens are required to be administered in adjuvant to elicit antibody production. For example, bovine serum albumin (BSA) in Complete Freund's adjuvant is immunogenic when injected into carp but is not when injected in saline unless the fish were first primed with this antigen in adjuvant.[28]

The immunogenicity of antigens varies between fish species. For example, BSA is a good immunogen in tilapia even without adjuvant[28] while it is a very poor immunogen in rainbow trout even when injected with adjuvant.[29] Horse erythrocytes are a good immunogen in the winter flounder but sheep erythrocytes are not.[30] However, sheep erythrocytes are excellent immunogens in the carp.[14,23]

Route of administration

Several authors have reported that intramuscular injection of antigen elicits somewhat higher antibody titres than intraperitoneal injection. Carp primed with sheep erythrocytes intramuscularly developed better memory than by the intravenous route.[23] Soluble BSA injected intracardially in carp induced tolerance to this antigen for over 16 months.[26]

Immersion immunization may or may not result in production of antibodies within the serum of carp following primary stimulation with *Aeromonas hydrophila*. However, on subsequent immersion, serum antibodies are produced indicating the establishment of memory which persisted for about 8 months. However, the stimulation of this memory component was not elicited by intramuscular injection of the antigen. Thus, the route of secondary stimulation had to be the same as the primary in order to activate the immune memory.[31]

Oral immunization does not usually elicit production of antibody in the serum but antibody may appear in the gut mucus. Oral immunization of trout with *Aeromonas salmonicida* actually suppressed serum antibody titres on subsequent injection of the antigen suggesting the induction of systemic suppressor cells after oral stimulation.[32] However, systemic suppression was not found in rainbow trout fed with a *Vibrio* vaccine[33] since agglutinating antibodies appeared in the serum 4 weeks following an oral vaccination (administered for 32 days).

These findings suggest that the immune system has somewhat separate parts – systemic and mucosal, the latter possibly being separated into integumentary and alimentary. The route of antigen stimulation may have important effects on any one of these parts. However, the effect on the immune system may be different for different antigens administered by different routes.

Adjuvants

Adjuvants are substances which enhance the immune response. They comprise a variety of substances, the mechanism of action being variable and in many cases not well understood. The humoral immune response, especially to T-dependent antigens can be considerably enhanced by administering the antigen with adjuvant. Protection of fish against *vibriosis* and ERM can be effectively achieved without the use of adjuvant and it may be significant that in both cases the protective antigens are lipopolysaccharide, considered to be a T-independent antigen. Protection against certain other diseases has only been achieved when antigen was given in adjuvant (e.g. *Aeromonas* vaccines,[34] BKD vaccines[35]). In these instances the adjuvant may either improve the immunogenicity of the antigens or have a non-specific stimulatory effect on non-specific defence mechanisms, or both (see below). In any case, the ineffectiveness of many fish vaccines when used without adjuvant has led to their inclusion in many vaccine trials. Unfortunately, some of the more effective adjuvants, frequently used in experimental procedures, are unacceptable in commercial practice because they cause severe local tissue responses and other side-effects.

Many adjuvants can, because of their physical nature, only be administered by injection. Some have the potential to be used in immersion or oral vaccination but to date very few trials have been conducted to assess their value in such methods.

Injectable adjuvants

The most effective adjuvant and the one most widely used by experimental immunologists is Freund's Complete Adjuvant, FCA, which is a mixture of killed *Mycobacteria tuberculosis* in a mineral oil in which the antigen in saline is emulsified. The emulsion can only be administered by injection. Unfortunately, FCA has many undesirable side-effects (including production of local granulomata, autoimmune diseases and tuberculin sensitization) and cannot be used in commercial vaccines for mammals. In fish it also induces side-effects. By intramuscular injection it induces sterile abscesses with extensive local muscle necrosis. Intraperitoneal injection is less traumatic but visceral adhesions and granulomata may result. FCA has been used in a number of field trials to vaccinate fish by intraperitoneal injection with significant enhancement of immune responses and protection. However, slight reductions in the growth rate of fish so treated may result.[36]

Freund's Incomplete Adjuvant, FIA, lacks the *Mycobacteria*. It is used in the same way as FCA, has only slightly less traumatic side-effects, but is not as effective an immunopotentiator as FCA. Other injectable adjuvants, which are used in commercial vaccines, are aluminium hydroxide gel mixed with saponin (in foot and mouth disease vaccine) and certain oily emulsions (of confidential composition e.g. in Newcastle disease vaccines for poultry). These adjuvants have had only limited trials in fish.[37] Aluminium hydroxide and saponin had no adjuvant effect in rainbow trout immunized with the protein-hapten antigen key-hole limpet haemocyanin-dinitrophenol, KLH–DNP, but the oily emulsions had some effect (about the same as FIA). Another alum adjuvant is potassium aluminium sulphate which has been used in an injectable *Vibrio* vaccine. It is not to be recommended as no enhanced protection was achieved and the adjuvant caused serious side-effects including peritonitis, depressed growth rate and, especially in fish of 3 cm length, substantial mortalities.[36]

Immersion and oral adjuvants

There are a number of adjuvants which theoretically allow their use with immersion or oral vaccines for fish but as yet very few have been tested in this way, though some have been tested by injection. Still others, which have adjuvant activity in mammals, have yet to be tested in fish by any route.

Potassium aluminium sulphate treatment of *Vibrio* immersion vaccine significantly increased the uptake of antigen[38] which would be expected to induce better protection. Protection resulting from oral administration of *Vibrio* vaccines has been considerably improved by treating the vaccine with this alum adjuvant and though the protection took longer to develop (8 weeks) as compared with injection vaccines (2 weeks), the degree of protection (70 per cent survival cf. 0 per cent in controls) though less than with injection administration, may be ample for commercial purposes.[33]

Dimethyl sulphoxide (DMSO) has been reported to increase antibody production to *Y. ruckeri* in rainbow trout when combined with the bacterium in immersion vaccination.[12]

A number of other adjuvants with potential use in immersion or oral vaccination but which have only been tested by injection include:

Muramyl dipeptide (MDP). This had no adjuvant activity in rainbow trout in terms of enhanced antibody production.[37] However, when administered in a mineral oil it did enhance non-specific protection mechanisms in coho salmon[39] (see Immunostimulants below).

Levamisole had similar, though less marked, effects as MDP in stimulating non-specific defence mechanisms.[39]

Ete is a soluble extract of the marine tunicate *Ecteinascidia turbinata* and enhances the antibody response of eels when injected 2 days following antigen injection.[40]

Other adjuvants which have proved useful for oral vaccines in mammals but which have not yet been tested in fish are cholera toxin[41] and *Quillaja* saponin.[42]

Immunostimulants

Certain substances, including some adjuvants, have the ability to increase resistance to infectious diseases, not by enhancing the specific immune response, but by enhancing non-specific defence mechanisms, particularly the phagocytic system. This resistance to infection is non-specific and therefore has no memory component and is likely to be of short duration. Nevertheless, these substances may have some value in the prophylaxis of fish diseases.

FK-156 is an immunoactive peptide which protects mice against microbial infection by both injection and oral administration. It has been shown to increase markedly the resistance of rainbow trout to *Aeromonas salmonicida* infection when injected 1 day prior to challenge. This effect was still marked but less when injected up to 7 days prior to challenge. The mechanism of action appears to be by increasing phagocyte activity and was still effective in immunosuppressed fish.[43]

Ete injection into eels enhanced non-specific resistance to a number of bacterial pathogens[44] including *A. hydrophila* infection.[45] Modified Freund's Complete Adjuvant (containing *Mycobacterium butyricum* instead of *M. tuberculosis*) caused a dramatic enhancement of non-specific resistance when injected into coho salmon. Marked resistance to experimental furunculosis and vibriosis was induced which persisted for at least 90 days.[39]

In conclusion, the non-specific defence mechanisms activated by certain adjuvants and immunostimulants may have practical application for aquaculture. If relatively long-term increase in resistance results, without accompanying undesirable side-effects, these substances may eliminate the need for certain vaccines which are too expensive to produce commercially. Further work to test the potency, duration and breadth of the protection induced by a larger number of immunostimulants and to investigate methods for their mass administration would be of extreme value.

REFERENCES

1 Ellis, A. E. (1981). Stress and the modulation of defence mechanisms in fish. In: *Stress and Fish*, ed. A. D. Pickering, London: Academic Press, pp. 147–169.
2 Pickering, A. D. and Pottinger, T. G. (1985). Cortisol can increase the susceptibility of brown trout, *Salmo trutta* L., to disease without reducing the white blood cell count. *J. Fish Biol.*, **27**, 611–619.
3 Ellsaesser, C. F. and Clem, L. W. (1986). Haematological and immunological changes in channel catfish stressed by handling and transport. *J. Fish Biol.*, **28**, 511–521.
4 Peters, G. and Hong, L. Q. (1985). Gill structure and blood electrolyte levels in European eels under stress. In: *Fish and Shellfish Pathology*, ed. A. E. Ellis, London: Academic Press, pp. 183–198.
5 Barrow, J. H. (1955). Social behaviour in fresh-water fish and its effect on resistance to trypanosomes. *Proc. Nat. Acad. Sci. USA*, **41**, 676–679.
6 Henderson-Arzapalo, A., Stickney, R. R. and Lavis, D. H. (1980). Immune hypersensitivity in intensively cultured *Tilapia* species. *Trans. Am. Fish. Soc.*, **109**, 244–247.
7 Schreck, C. B. (1981). Stress and compensation in teleostean fishes: Response to social and physical factors. In: *Stress and Fish*, ed. A. D. Pickering, London: Academic Press, pp. 295–321.
8 Durve, V. S. and Lovell, R. T. (1982). Vitamin C and disease resistance in channel catfish (*Ictalurus punctatus*). *Can. J. Fish. Aquatic Sci.*, **39**, 948–951.
9 Li, Y. and Lovell, R. T. (1985). Elevated levels of dietary ascorbic acid increase immune responses in channel catfish. *J. Nutrition*, **115**, 123–131.
10 Blazer, V. S. and Wolke, R. E. (1984). Effect of diet on the immune response of rainbow trout (*Salmo gairdneri*). *Can. J. Fish. Aquatic Sci.*, **41**, 1244–1247.
11 Cowey, C. B., Adron, J. W. and Youngson, A. (1983). The Vitamin E requirement of rainbow trout (*Salmo gairdneri*) given diets containing polyunsaturated fatty acids derived from fish oil. *Aquaculture* **30**, 89–93.
12 Anderson, D. P., Van Muiswinkel, W. B. and Roberson, B. S. (1984). Effects of chemically induced immune modulation on infectious diseases of fish. In: *Chemical Regulation of Immunity in Veterinary Medicine*, ed. M. Kende, J. Gainer and M. Chirigos, New York: A. L. Liss, pp. 182–211.
13 O'Neill, J. G. (1981). Effects of intraperitoneal lead and cadmium on the humoral immune response of *Salmo trutta*. *Bull. Environ. Contam. Toxicol.*, **27**, 42.
14 Rijkers, G. T., Teunissen, A. G., Van Oosterom, R. and Van Muiswinkel, W. B. (1980). The immune system of cyprinid fish. The immunosuppressive effect of the antibiotic oxytetracycline in carp (*Cyprinus carpio* L.). *Aquaculture*, **19**, 177–189.

15 Van Muiswinkel, W. B., Anderson, D. P., Lamers, C. H. J., *et al.* (1985). Fish immunology and fish health. In: *Fish Immunology*, ed. M. J. Manning and M. F. Tatner, London: Academic Press, pp. 1–8.
16 Grondel, J. L., Nouws, J. F. M., and Van Muiswinkel, W. B. (1987). The influence of antibiotics on the immune system: immuno-pharmokinetic investigations on the primary anti-SRBC response in carp, *Cyprinus carpio* L., after oxytetracycline injection. *J. Fish Dis.*, **10**, 35–43.
17 Thorburn, M. A., Carpenter, T. E. and Ljungberg, O. (1987). Effects of immersion in live *Vibrio anguillarum* and simultaneous oxytetracycline treatment on protection of vaccinated and non-vaccinated rainbow trout, *Salmo gairdneri* against vibriosis. *Dis. Aq. Organisms*, **2**, 167–171.
18 Avtalion, R. R. (1981). Environmental control of the immune response in fish. *Crit. Rev. Environ. Control*, **11**, 163–188.
19 Bly, J. E., Buttke, T. M., Meydrech, E. F. and Clem, L. W. (1986). The effects of *in vivo* acclimation temperature on the fatty acid composition of channel catfish (*Ictalurus punctatus*) peripheral blood cells. *Comp. Biochem. Physiol.*, **83B**, 791–795.
20 Horne, M. T., Tatner, M., McDerment, S., Agius, C. and Ward, P. (1982). Vaccination of rainbow trout, *Salmo gairdneri* Richardson, at low temperatures and the long persistence of protection. *J. Fish Dis.*, **5**, 343–345.
21 Tatner, M. F. and Horne, M. T. (1984). The effects of early exposure to *Vibrio anguillarum* vaccine on the immune response of the fry of the rainbow trout, *Salmo gairdneri* Richardson. *Aquaculture*, **41**, 193–202.
22 Bly, J. E., Cuchens, M. A. and Clem, L. W. (1986). Temperature-mediated processes in teleost immunity: binding and mitogenic properties of concanavalin A with channel catfish lymphocytes. *Immunology*, **58**, 523–526.
23 Rijkers, G. T., Frederix-Walters, E. M. H. and Van Muiswinkel, W. B. (1980). The immune system of cyprinid fish. Kinetics and temperature dependance of antibody producing cells in carp (*Cyprinus carpio*). *Immunology*, **41**, 91–97.
24 Zeeman, M. G. (1986). Modulation of the immune response in fish. *Vet. Immunol. Immunopathol.*, **12**, 235–241.
25 Nakanishi, T. (1986). Seasonal change in the humoral immune response and the lymphoid tissues of the marine teleost, *Sebastiscus marmoratus*. *Vet. Immunol. Immunopathol.*, **12**, 213–221.
26 Wishkowsky, A. and Avtalion, R. R. (1982). Induction of helper and suppressor functions in carp (*Cyprinus carpio*) and their possible implications in seasonal disease in fish. *Dev. Comp. Immunol., Suppl.* **2**, 83–91.
27 Lamers, C. H. J., De Haas, M. J. H. and Van Muiswinkel, W. B. (1985). Humoral response and memory formation in carp after injection of *Aeromonas hydrophila* bacteria. *Dev. Comp. Immunol.* **9**, 65–75.
28 Avtalion, R. R., Wishkowsky, A. and Katz, D. (1980). Regulatory effect of temperature on specific suppression and enhancement of the humoral immune response in fish. In: *Phylogeny of Immunological Memory*, ed. M. J. Manning, Amsterdam: Elsevier/North Holland, pp. 113–121.
29 Hodgins, H. O., Weiser, R. S. and Ridgeway, G. J. (1967). The nature of antibodies and the immune response in rainbow trout (*Salmo gairdneri*). *J. Immunol.*, **99**, 534–544.
30 Stolen, J. S., Gahn, T. and Nagle, J. (1982). The humoral antibody formation to erythrocyte antigens in three species of flatfish. *Dev. Comp. Immunol., Suppl.* **2**, 101–106.

31 Lamers, C. H. J., De Haas, M. J. H. and Van Muiswinkel, W. B. (1985). The reaction of the immune system of fish to vaccination: development of immunological memory in carp (*Cyprinus carpio*) following direct immersion in *Aeromonas hydrophila* bacterin. *J. Fish Dis.*, **8**, 253–262.
32 Udey, L. R. and Fryer, J. L. (1978). Immunization of fish with bacterins of *Aeromonas salmonicida*. *Mar. Fish. Rev.*, **40**, 12–17.
33 Agius, C., Horne, M. T. and Ward, P. D. (1983). Immunization of rainbow trout, *Salmo gairdneri* Richardson, against vibriosis: comparison of an extract antigen with whole cell bacterins by oral and intraperitoneal routes. *J. Fish Dis.*, **6**, 129–134.
34 Paterson, W. D. (1981). *Aeromonas salmonicida* as an immunogen. *Develop. Biol. Standard.*, **49**, 375–385.
35 Klontz, G. W. (1983). Bacterial kidney disease in salmonids: an overview. In: *Antigens of Fish Pathogens*, ed. D. P. Anderson, M. M. Dorson and P. Dubourget, Lyon: Fondation Marcel Merieux, pp. 177–198.
36 Horne, M. T., Roberts, R. J., Tatner, M. and Ward, P. (1984). The effects of the use of potassium alum adjuvant in vaccines against vibriosis in rainbow trout, *Salmo gairdneri* Richardson. *J. Fish Dis.*, **7**, 91–99.
37 Cossarini-Dunier, M. (1985). Effect of different adjuvants on the humoral immune response of rainbow trout. *Dev. Comp. Immunol.*, **9**, 141–146.
38 Tatner, M. F. and Horne, M. T. (1983). Factors influencing the uptake of ^{14}C-labelled *Vibrio anguillarum* vaccine in direct immersion experiments with rainbow trout, *Salmo gairdneri* Richardson. *J. Fish Biol.*, **22**, 585–591.
39 Olivier, G., Evelyn, T. P. T. and Lallier, R. (1985). Immunity to *Aeromonas salmonicida* in coho salmon (*Oncorhynchus kisutch*) induced by modified Freund's complete adjuvant: its nonspecific nature and the probable role of macrophages in the phenomenon. *Dev. Comp. Immunol.* **9**, 419–432.
40 McCumber, L. J., Trauger, R. and Sigel, M. M. (1981). Modification of the immune system of the American eel, *Anquilla rostrata* by ETE. *Develop. Biol. Standard.*, **49**, 289–294.
41 Lycke, N. and Holmgren, J. (1986). Strong adjuvant properties of cholera toxin on gut mucosal immune responses to orally presented antigens. *Immunology*, **59**, 301–308.
42 Maharaj, I., Froh, K. J. and Campbell, J. B. (1986). Immune responses of mice to inactivated rabies vaccine administered orally: potentiation by *Quillaja* saponin. *Can. J. Microbiol.*, **32**, 414–420.
43 Kitao, T. and Yoshida, Y. (1986). Effect of an immunopotentiator on *Aeromonas salmonicida* infection in rainbow trout (*Salmo gairdneri*). *Vet. Immunol. Immunopathol.*, **12**, 287–296.
44 Sigel, M. M., McCumber, L. J., Hightower, J. A., *et al.* (1983). *Ecteinascidia turbinata* extract activates components of inflammatory responses throughout the phylogenetic spectrum. *Amer. Zool.*, **23**, 221.
45 Davis, J. F. and Hayasaka, S. S. (1984). The enhancement of resistance of the American eel, *Anguilla rostrata*, to the pathogenic bacterium *Aeromonas hydrophila* by an extract of the tunicate *Ecteinascidia turbinata*. *J. Fish Dis.*, **7**, 311–316.
46 Zeeman, M. G. and Brindley, W. A. (1981). Effects of toxic agents upon fish immune system. In: *Immunologic Considerations in Toxicology, Vol III*, ed. R. P. Sharma, Boca Raton, Florida: CRC Press, pp. 1–60.

4

Commercial Production and Licensing of Vaccines for Fish

M. T. Horne and A. E. Ellis

INTRODUCTION

Of the many infectious diseases that affect fish there are only two for which effective commercial vaccines have been produced i.e. ERM and vibriosis, including cold water vibriosis. This reflects the wide gap which exists between the demonstration of protection under controlled, experimental conditions and the manufacture of a commercial product which performs consistently to closely defined standards in commercial aquaculture. Furthermore, the final stages of development involved in large scale production, testing and licensing a vaccine for commercial use are lengthy and expensive. Although aquaculture is growing rapidly into a significant industry worldwide, at present few fish diseases are sufficiently serious in purely commercial, economic terms to warrant the scale of expenditure required to develop and license a commercial vaccine. In this chapter the broad aspects of commercial production and licensing of vaccines will be described.

PRODUCTION OF VACCINES

Identification of protective antigens

Early research is frequently centred around a small number of isolates of the pathogen, chosen primarily because antigen preparations from these are effective in conferring protection in experimental trials. Although the original vibriosis and ERM vaccines were developed without knowledge of the protective antigens, the development of vaccines to other fish diseases will almost certainly need to identify the protective antigens so that methods of optimizing their expression under mass production conditions can be developed.

More than one antigen may be protective, and some antigens may stimulate some parts of the immune system more than others so that the

FISH VACCINATION
ISBN 0-12-237485-1

selection of strains for oral, immersion and injected vaccines potentially, might differ.

When a certain antigen, such as the lipopolysaccharide (LPS) of *Vibrio anguillarum* and *Yersinia ruckeri* is identified as the prime protective component over a wide range of isolates (see Ch. 6 and 7), it is then necessary to investigate geographic and host species variation before formulating a vaccine.

Determination of antigens is most readily accomplished by using a combination of gel electrophoresis and blotting techniques. Suitable preparation of the pathogen's antigens (e.g. whole cell lysates, outer membrane components, extracellular products) are separated, usually by electrophoresis and the subsequent banding patterns studied. Transfer to nitrocellulose membranes by electroblotting enables the preparation to be handled more effectively in the subsequent staining and ELISA procedures used to identify the antigenic components.

The binding of antibody to the separated antigens is detected by using a suitable 'anti-fish antibody' conjugated to a radioactive or enzyme tracer or avidin–biotin system. The number of bands recognized by the animal as antigenic may differ in different species. Thus, Hastings (see Ch. 8) showed that, whereas rabbits may recognize 25 components of *Aeromonas salmonicida* extracellular antigens, only four or five are recognized by rainbow trout. Little information comparing different fish species has yet been obtained but from the example of mammalian systems, it is likely fish from different genera and species will differ in respect of the antigens which they recognize.

The gel-blotting–antibody probe systems described can also be used quantitatively to determine the amount of a particular antigen present in different strains but there is no direct way of identifying which antigens are protective. This is accomplished by further experimentation or comparison of a library of strains for which the antigen spectrum and protective ability is known. If protection is based upon antibodies, it is possible to identify protective antigens by passive immunization using a selection of monospecific antibodies directed to different antigens considered likely to be protective. However, furunculosis and BKD are examples of important fish diseases where an effective specific-immune response is not induced in the fish host and the pathogens are probably responsible for a degree of immune repression.[1] Empirical vaccine formulations are not effective under these circumstances and different basic methodologies need to be adopted. For example, it may be possible to identify the protective antigens by passive immunization with antibodies raised in a responsive species. Using such a procedure, Ellis *et al.*[2] (see Ch. 8) identified the extracellular protease of *A. salmonicida* as a protective antigen. Methods of improving the immunogenicity of this antigen need then to be sought to induce a protective response in the fish host.

Source of protective antigens: serotypes and strains

Species of bacterial and viral pathogens often show variation in serotype groupings. This is because of variation in certain surface antigens which are strongly immunogenic and is commonly assessed by simple slide agglutination tests using rabbit antiserum. In bacteria, serotype differences are frequently associated with variation in the structure of the LPS. If antigens which represent different pathogenic groups or strains are also involved in eliciting a protective immune response (as with vibriosis and ERM vaccines) then the different serotypes must be incorporated into the vaccine to protect against the different serotypes of the pathogen. Such a vaccine is termed multivalent.

Existing commercial vaccines for fish are of the simplest type and because the target diseases have a global occurrence there has been initially a tendency to promote vaccines developed in one region in any area where apparently similar disease occurs. Generally this has been successful and probably the strong stimulation of non-specific defence mechanisms by *Vibrio anguillarum* antigens may have contributed to this. However, it is becoming clear that significant serotype variation does occur and that adjustments of vaccine composition to regional strains may be beneficial or necessary.

Recent work on *Yersinia ruckeri* shows that this species is variable and that European, British and American strains are probably different. The LPS serotypes I, II and V have been implicated in disease although the majority of vaccines are monovalent (composed only of serotype I). Commercial *Vibrio* vaccines are usually bivalent (composed of two serotypes) although trivalent experimental types (also including *V. salmonicida*) have been used. It should be noted that although manufacturers frequently cite bivalent *Vibrio* vaccines to contain both *V. anguillarum* and *V. ordalii* the biochemistry used to define these two species is independent of the division into the two main LPS serotypes. *V. ordalii* expresses only one serotype (2 or J-0-1) while *V. anguillarum* is heterogeneous with up to three serotypes including type 2.[3] It is therefore irrelevant which of the two species is used to provide the serotype 2 (J-0-1) component.

An early decision in vaccine development is therefore the serologically determined, strain composition of the vaccine and whether fish species or geographically correlated subtypes are required. However, it is important to remember that the immunodominant components which may or may not be associated with serotype identity may not be the protective antigens (as in *Aeromonas*). In studying regional serotypic variations from a purely descriptive point of view this is not a problem, but in determining suitable strains for vaccine formulation it is critical. Only when antigens responsible for serotype identity are also protective, and this coincidence is not inevitable, does it become necessary for a vaccine to contain the appropriate serotype antigens.

Manufacture and stability of the protective antigens

Further factors in strain selection, of considerable commercial importance but rarely considered by experimental laboratories, are those of viability and antigen expression following long-term storage. Early in the development of a vaccine, strains identified for possible inclusion are stored by a number of routine methods (freeze dried, liquid nitrogen etc.) and periodically reassayed for viability and antigen expression. The significant protective antigens of both *Yersinia* and *Vibrio* (except *V. salmonicida*) are stable in freeze-dried and cryopreserved samples for extended periods but most pathogenic bacterial species show significant change in gene expression during storage and sometimes these may be restored only by passaging in the host species.[4] If these properties are key factors in inducing protective immunity then special care has to be taken in storage or alternative methods of retaining viable strains adopted.

Strains chosen for production, in addition to storage stability, must be sufficiently robust to survive mass culture and further research may be necessary to determine the point of the growth cycle or the most effective growth medium to maximize protective antigen expression. Where expression predominates over a narrow range of the growth cycle, particularly if this point occurs early, in 'young' cultures, it may be necessary either to concentrate the culture to obtain sufficent cell density, or to resort to continuous culture and harvesting.

Not all potentially useful strains survive the vigorous aeration and agitation which can occur in commercial culture without extensive lysis. If other, more robust strains with suitable antigenic properties are not available then alternative conditions, usually less optimal, economically, need to be substituted.

None of the potential problems of culturing useful but 'difficult' strains is without a solution but any departure from the simple, routine methods increases cost and, in the case of fish vaccines, production at low cost is always likely to be a dominating factor in view of the relatively small value of each market. The opportunities becoming available through genetic engineering, where it may well be possible to move the genes expressing the protective antigen to a more robust and, in the case of viruses, a bacterial host, could eventually make more fish vaccines available cost effectively.

LICENSING VACCINES

Once the techniques for mass production of a vaccine have been developed, a licence must be obtained to market the product and this in itself may take several years of testing and development. Licensing authorities require that a commercial vaccine should perform consistently within given parameters. Even with a reliable organism grown under a 'simple' regime, variation can occur between culture batches. Licensing requirements vary considerably in different countries but typically a test protocol outlining the acceptable performance limits of any vaccine batch released is agreed between a manufacturer and the authorities. Data is usually required from three early batches demonstrating consistency in that the set criteria are met by all three, before a licence application is accepted. The set criteria are: potency, safety and stability ('shelf-life').

Potency testing

Each batch of vaccine must meet a minimum standard in potency tests for approval. Fish are vaccinated and challenged (by immersion or injection) and the mortalities compared with unvaccinated control fish. The details of the methodology and handling of the data for calculating potency of vaccines in fish were developed by Amend.[5] The method allows the calculation of a value of potency expressed as the Relative Per cent Survival (RPS) (see Ch. 1 for further details). Most licensing authorities have broadly accepted these or similar criteria for potency testing and within the defined limits of the criteria a minimum RPS value must be obtained for a vaccine to be approved. However, these criteria were originally developed for potency testing of *Vibrio* vaccines and it may not always be possible in practice, to achieve results from tests with other vaccines, which exactly meet the limitations set by Amend.[5] For example, it may be difficult with some diseases to consistently reach the recommended mortality rates (60 per cent) in control unvaccinated fish during experimental challenges. Thus, it will probably be necessary for licensing authorities to formulate criteria of acceptability which are appropriate for each disease. Most licensing authorities require the tests to be performed on the species of fish for which the vaccine is recommended for use by the manufacturer. Data on the duration of protection may also be required and generally this must reflect the needs of the farm husbandry cycles and the commercial market for which fish are being reared. Protection against vibriosis is usually high for at least 12 months; protection against ERM declines more quickly.

The choice of a challenge strain to establish potency is clearly important and either a standard virulent strain is registered or a recent field isolate, confirmed and certified to be of the correct serotype, is used and notified. In either case it is advisable to confirm virulence at the experimental temperature by a trial experiment before batch testing. Part of this procedure ensures that 'master seed' cultures for both challenge and production are stored properly and protocols for the manner and frequency of subculturing are clearly defined and agreed. Both the methodology and the site of batch testing are required to be approved. Tests are therefore usually carried out in establishments to which an independent auditing body has access to ensure that 'Good Laboratory Practice' is maintained and records at all levels of processing from collection to final presentation, are available for scrutiny.

The above concerns laboratory or pilot-scale tests used in vaccine development. However, licensing authorities also require that evidence of potency under field conditions should be presented. Frequently this is required in the country in which the application is lodged independently of data obtained elsewhere in the world or whether the vaccine is already licensed for use in some other country.

Authorities are aware of the difficulties of carrying out rigorous experimentation under farm conditions with natural challenge and for this reason no rigid performance criteria are set. It is required minimally that there should be clear evidence that a natural challenge has occurred and that the level of specific mortality in the vaccinates is unambiguously lower than in non-vaccinated controls. Occasionally, where all fish are vaccinated, performance equal to or better than an existing licensed vaccine, used in the same trial, may be considered. Despite the difficulties involved most field trials usually provide very much more data, including the level of non-specific mortalities, feed conversion rate comparisons etc, which are invaluable in offering guidance for the use of vaccines by the farmer.

Safety

The foregoing is largely concerned with consistency of performance so that a marketed product can substantiate the claims made for it. However, of primary concern to the user is the question of safety. All commercial, bacterial vaccines used in aquaculture at present are inactivated by the addition of 0.5 per cent or more formalin followed by a viability test 24 or 48 h later to confirm sterility. Some manufacturers additionally use a preservative such as thiomersalate, to further protect the culture from degradation. Addition of formalin can denature antigens and higher concentrations (e.g. 3 per cent) are frequently used to inactivate toxins in vaccines although this may also alter antigenicity.

Toxicity is particularly important and, under certain conditions, *Vibrio, Yersinia* and *Aeromonas* vaccine preparations have proved toxic to fish. It is usually required in testing that a 'double-safe' procedure is employed whereby the fish are exposed to twice the concentration recommended for use. So, for example, where a 1:10 dilution of a vaccine is recommended for use it is implied that no adverse effects were seen in tests previously carried out at 1:5; similarly a 1:500 bath vaccine will have been tested for 1 h at 1:250.

It is particularly important when recommending use in a 'new' species that toxicity data should be obtained as species are differentially sensitive to all commercial vaccines presently in use. Rainbow trout in sea or fresh water will survive indefinitely in strong concentrations of *Vibrio* bacterin whereas sea-bass are particularly sensitive (Horne, unpublished). Temperature can also influence toxicity and the temperature range or values for the test must be defined and agreed.

Shelf-life and stability

The type of container in which the vaccine is to be marketed, although viewed primarily as a marketing decision, needs to be finalized at an early stage of development as the type of glass or plastic used can affect the stability of the antigen. For the purpose of licensing a vaccine the shelf-life and stability must therefore be determined in the type of container in which it will be sold. Formalized bacterins of the type used in most commercial fish vaccines, particularly where LPS is the immunogen, are usually very stable and have been shown to have substantial potency for as long as 5 years. Normally under UK legislation a nominal shelf-life is agreed with a manufacturer for test purposes but the statement of shelf-life on a production-batch label must be established by experimentation and the statement can be no more than 6 months less than the actual time tested. To obtain a shelf-life value which can be stamped on all new batches, it is necessary to test three similar batches, stored as recommended to the user, over an extended period. The majority of vaccines currently available have a shelf-life of 3 years indicating therefore that performance was maintained in three batches over at least 3.5 years. Any decision to modify the container in which the vaccine is supplied must bear in mind that stability data may need to be obtained in the new container if the material of construction is different.

CONCLUSIONS

This chapter describes the scientific and licensing procedures involved in taking a successful vaccine from the small experimental scale of the laboratory to the stage of commercial markets. Even for 'simple' vaccines, such as the present vibriosis and ERM vaccines, this procedure is expensive and lengthy. At present there are only a few fish diseases which are serious enough to the aquaculture industry to provide a manufacturer of commercial vaccines sufficient capital return to merit the investment required for marketing an appropriate vaccine. With continued rapid growth of fish farming worldwide the potential for vaccines will also increase and the new biotechnologies of genetic and protein engineering may provide manufacturers with more effective techniques to mass produce other, particularly viral, vaccines to fish diseases where traditional methods of antigen production and purification at the present are too expensive.

REFERENCES

1 Turaga, P., Weins, G. and Kaattari, S. (1987). Bacterial kidney disease; the potential role of soluble protein antigen(s). *J. Fish Biol.*, **31**, (Suppl. A) 191–194.
2 Ellis, A. E., Burrows, A. S., Hastings, T. S. and Stapleton, K. J. (1988). Identification of *Aeromonas salmonicida* extracellular protease as a protective antigen against furunculosis by passive immunisation. *Aquaculture* **70**, 207–218.
3 Horne, M. T. (1982). The pathogenicity of *Vibrio anguillarum* (Bergman). In: *Microbial Diseases of Fish*, ed. R. J. Roberts, London: Academic Press, pp. 171–187.
4 Adams, A., Bundy, A., Thompson, K. and Horne, M. T. (1988). Association between virulence and cell surface characteristics. *Aquaculture* **69**, 1–14.
5 Amend, D. F. (1981). Potency testing of fish vaccines. *Dev. Biol. Standard*, **49**, 447–454.

5

Strategies of Fish Vaccination

M. T. Horne and A. E. Ellis

INTRODUCTION

In relation to other methods of disease control i.e. exclusion and chemotherapy, the epidemiological aspects of vaccination strategies were considered in Chapter 1. In this chapter two other strategic factors will be discussed: first, the strategic use of vaccination methods i.e. injection, immersion and oral, in relation to the size and biology of the fish and the nature of the farm (intensive or extensive); secondly, the economic factors underlying vaccination strategies and the important parameters which need to be built into models for carrying out a cost-benefit analysis.

STRATEGIC USE OF VACCINATION METHODS

Different farm circumstances have given rise to a number of different methods of administering vaccines. The basic methodologies adopted are injection, immersion (including spraying) and oral. Injection and immersion methods are suitable only for intensive aquaculture and both require the fish to be handled or at least confined in a small space during the procedures. Oral vaccination is the only method economically suited to extensive aquaculture, but of course, can be administered only in an artificial diet. The relative advantages, disadvantages and applications are summarized in Tables 5.1 and 5.2.

Injection

It is essential for injection vaccination that fish are anaesthetized not merely to facilitate handling but to avoid excessive stress. The two commonest anaesthetics in general use are MS222 and benzocaine. MS222 is directly soluble in water and light anaesthesia is achieved at 50 mg/litre (50 ppm). However, it is costly for extensive use. Benzocaine is dissolved in the minimum volume of alcohol before addition to water at a concentration of approx 25 mg/litre. In a typical farm scale vaccination, fish which have fasted for 24 h are crowded to one end of the holding cage or pond and small

FISH VACCINATION
ISBN 0-12-237485-1

netsful lifted for immersion in an anaesthetic bath. Immediately the fish roll on their sides they are drained in the net and transferred to polystyrene trays which minimize damage. This arrangement is suitable for sea cages where small tables may be secured on the walkways servicing one or two operators. Alternatively, they may be passed down an irrigated chute to a group of workers around a vaccinating table, a method particularly suited to larger scale work on a land-based tank farm site. In both cases fish are presented 'dry' to the injectors which considerably speeds handling and makes possible long working periods without the hands being in water or anaesthetic. The vaccine is injected using multidose syringes.

Fish are presented ventrally to the needle with the head facing away from the operator so that the needle slides easily beneath the scales. The point of insertion is immediately anterior to the pelvic fins, with the syringe held at an angle of about 25° to the body surface so that the inoculum enters the peritoneal cavity with a minimum risk of damaging the underlying organs. Needles should be changed as soon as they become blunted. Commercially available vaccines are mostly formulated for a dose rate of 0.1 ml independently of fish size.

Table 5.1 *A summary of advantages and limitations of injection, immersion and oral vaccination methods*

Method	Advantages	Limitations
Injection	Most potent immunization route. Allows use of adjuvants. Most cost effective method for large fish.	Useful only in intensive aquaculture. Labour intensive. Stressful (anaesthesia; handling). Fish must be >15 g.
Immersion	Allows mass vaccination of small (<5 g) fish. Most cost effective method for fish <10 g. Bath: not stressful.	Only for intensive aquaculture. Dip: handling stress. Potency not as high as injection route.
Oral	Only method for extensive aquaculture. Non-stressful. Allows mass vaccination of fish any size. No extra labour costs.	Poor potency. Requires large amounts of vaccines to achieve protection. Suitable only for fish fed artificial diet.

Table 5.2 *Method of administering vaccines to fish*

Route	Vaccine dilution*	Exposure time
Immersion dip	1 : 3/1 : 10/1 : 100	5–30 s
Immersion bath	1 : 500/1 : 5000	1 h or longer
Spray	1 : 3/1 : 10/1 : 100	2–5 s
Inoculation	undiluted (0.1 ml)	Direct injection
Oral	in food	1 week or longer

*Dependent on type of vaccine and size of fish

Intramuscular injection on the dorsal surface of the fish is rarely used mainly due to the possibility of creating unsightly scarring but also because of the leakage of the inoculum before adsorption. If adjuvants (e.g. oil or alum based) are present in the vaccine, intramuscular injection should not be used as it causes extensive muscle necrosis which often ulcerates. Despite the presence of peritoneal pores in some fish species (e.g. salmonids) loss of vaccine from intraperitoneal injection is insignificant.

The rate achieved depends on the individual weight of the fish, the skill of the operator and the efficiency of the team of anaesthetists and fish handlers. Over a sustained period an experienced handler will inject 1000 to 1200 fish per hour when conditions are favourable and the fish are in excess of 20 g. The rate achieved rarely falls below 600–700 per hour even under poor conditions with inexperienced workers.

Handling of the fish is the critical part of the operation and the handlers are the key team members in ensuring that no losses occur. It is important that fish are anaesthetized lightly for a minimum time. Simultaneously fish must be presented to the injectors at a sufficient rate to ensure that they are able to work continuously at maximum speed. In balancing these objectives the fish condition is paramount and if a delay occurs it is the responsibility of the handlers to monitor the situation, if necessary returning uninjected fish to the holding pond. Providing care and commonsense are exercised no losses should result from the injection procedures if the fish are healthy at the start.

Injection is an advantageous mode of administration both from the point of view of the protection parameters and economics (see below). However, it is labour intensive, relatively slow and becomes impractical when fish are below 15 g. For experimental purposes even the smallest fry may be injected but in the farm situation the speed of operation falls sharply. Salmon smolts, which fall into the range of 25–45 g can be safely injected by experienced teams, but care is essential in carrying out these procedures at such a delicate stage of the salmon's life cycle and few salmon farmers will consider this procedure in the 4 weeks prior to smolting when substantial losses can occur even with careful handling.

However, once fully acclimatized in sea water, in tanks or cages, salmon are robust and, as size increases, other methods of vaccination become less economic (see below) and often from the handling point of view, impracticable. Injection, therefore, is predominantly used in cage situations or when large fish (in excess of 100 g) require to be treated as is the case when large trout are transferred to sea water for ongrowing to smoking size.

Immersion methods

All other methods in routine use expose the outer surface of the fish directly to diluted vaccine (direct immersion) relying on uptake, predominantly by the gills[1] and to some extent by imbibition[2,3] to take place passively. There are two variations of direct immersion; dip and bath vaccination.

Dip vaccination

A small tank containing vaccine is placed in close proximity to the holding tanks. The volume of the vaccine bath will depend on the manufacturer's instructions but most commonly vaccine is diluted one part vaccine to nine of clean water at the same temperature as the fish. Each litre of undiluted vaccine contains sufficient antigen for 100 kg of fish. Therefore for each 100 000 fish at 4.5 g (approx 100 per lb) being vaccinated, 4.5 litres of vaccine will be required giving a working volume of 45 litres when diluted. Table 5.3 shows the vaccine requirement for different fish sizes under this protocol.

The water level holding the first batch of fish is lowered to facilitate catching and small netsful of fish are taken, drained (to avoid diluting the vaccine) and still retained in the net, immersed in the vaccine for 20 s to 1 min according to the specific manufacturer's recommendation. The net of fish is then drained of vaccine and the fish returned to a fresh holding tank. It is advantageous to oxygenate the vaccine bath, particularly where a vaccine recommending a long exposure is used in warm weather. A specific recommendation for the amount of fish per netted batch is difficult to determine. However, it is important to realize that the fish should not be crowded if all are to receive a proper exposure. Typical recommendations of density of fish in the vaccine are 450 g of fish/litre of diluted vaccine.

The handling procedures, crowding and netting, required by dip vaccination are potentially stressful and may cause scale loss and a temporary lowering of fitness. For this reason vaccination is frequently carried out concurrently with grading where this is practised. However, even relatively delicate species such as sea bass and sea bream, carefully handled, can be treated this way.

Bath vaccination

The requirement for vaccination without netting or handling has led to the increasing use of the 'bathing' methods in which vaccine is added directly to the holding tank. The simplicity of this administration however, is counterbalanced by the much greater amount of vaccine required.

Addition of one part in 10 to a holding tank would require an excessive volume of vaccine. However, a more dilute solution requires the exposure time to be increased if the amount of vaccine taken up is to be sufficient to ensure long lasting immunity.

The 'standard' protocol for bath vaccination was developed by Egidius and Andersen[4] in Norway and is widely used for the administration of vaccine to salmon at the pre-smolt stage 4–8 weeks before transfer to sea water.

The water level of the holding tank is lowered to the minimum level in which it is judged the weight of fish in the tank can safely be held, with oxygenation, for 1 h. Vaccine is added to this volume to achieve a dilution of one part in 500. After 1 h the water flow is restored and the tank volume brought back to normal with the vaccine slowly diluting away. As in other methods the condition of the fish needs to be monitored with great care, particularly if temperatures are high, and the rate of oxygenation increased or water supply restored at any indication of excessive stress.

The weight of fish per litre of undiluted vaccine treated by this method varies according to farm circumstances where the holding density per tank and the manager's judgement of the minimum level are the two important factors. On average 40–60 kg of fish may be vaccinated per litre of vaccine but, where the holding density is low and it is not desirable to handle the fish in order to combine the stock of two tanks, as little as 25–30 kg fish/litre vaccine may still be cost effective especially with high value fish such as salmon smolts and turbot. Furthermore, this method is not labour intensive and one or two individuals can cover an entire farm in one morning.

It should be noted that immersion vaccines in high concentration can be toxic and it is vital to follow the manufacturer's recommendations. Moreover, fish of different species may be differentially sensitive to vaccines so it is important that when a vaccine is used on a species for which a protocol has not been tested by the manufacturer the farmer should perform an on-site test with a small batch of fish before vaccinating the main batch of stock. Additionally the volume of water at a given depth in the tank type used must be known and not guessed.

These precautions are particularly vital when a variation of this method is used which exposes the fish to a higher concentration of vaccine for part of the application period. When bath vaccinating fish in many tanks simultaneously, the limiting factor on a particular farm may be the availability of oxygenation equipment if all of the tanks are not permanently piped to a central supply. To overcome this, the volume of water for vaccination may be lowered to half of the level considered safe for 1 h, the vaccine added at 1:250 and the water input retained at a flow which will adjust the volume to the 'safe' level after 1 h. This obviates the need for oxygenation and results in a longer exposure per concentration factor for the fish. However, the potential dangers of this procedure must be borne in mind and the fish monitored constantly throughout. The fish initially will be exposed to double the concentration recommended by the manufacturer (although the vaccine will have been proved in development at this dose) and the fish are in less water than would be safe with no incoming, fresh supply. Although this method has been used successfully with salmon and trout the onus is strongly on the farmer or veterinarian firstly to ensure that, at the temperature of the administration, there is no toxic effect by carrying out a preliminary trial at least 24 h previously and secondly to adjust the water supply as required if fish behaviour suggests stress during the administration period.

For Atlantic salmon, ERM vaccines are more toxic than *Vibrio* vaccines while rainbow trout are insensitive to both under normal growth conditions. Similarly, whereas rainbow trout will survive indefinitely in strong concentrations of *Vibrio* bacterin in sea or fresh water, sea bass are particularly sensitive (Horne, unpublished).

Theoretically it is possible to use either the dip or bathing methods in the sea cage situation, but in both cases administration is probably uneconomic. It has been explained that the antigen concentration is fixed at a given weight of fish per litre so that relatively few doses are achieved when fish of typical cage weights are treated, even though it is possible to crowd, net and dip cage fish fairly easily. If a tarpaulin is passed inside the net to create a pond the administration is even less economic since the minimum volumes of water obtained are so large that an excessive amount of vaccine is required.

Oral vaccination

To many fish farmers, merely adding antigen to the food for a suitable time course, is compellingly simple and attractive. For extensive aquaculture where fish are farmed in very large ponds and handled only when harvested, oral vaccines are the only practical strategy possible. The earliest attempts at fish vaccination[5] used oral administration and much research has been devoted towards this end for all of the bacterial vaccines currently in use. It is widely held that protection achieved by this method takes longer to achieve and is lower than with the other methods (see also Ch. 1). However, careful scrutiny of the literature shows that this is far from clear cut and that when sufficient antigen mass is administered protection may be acceptably high. Furthermore, potassium alum has been shown to markedly increase the potency of oral *Vibrio* vaccines[6] and further research into the use of adjuvants with oral vaccines is certainly warranted (see also Ch. 3).

Automation of vaccination

Automation of vaccination has been little used but the immersion methods do lend themselves to mechanization. Although efficient and acceptable in their labour requirements immersion methods are not always easily adaptable to particular farm circumstances where very large numbers of fish may create some difficulties. Automation was first attempted by Gould *et al.*[7] using a spray application which has become extensively used on large farm units in the USA. Fish are drawn onto a conveyor belt and passed beneath a spray set to ensure at least a 5 s exposure to the diluted vaccine. The vaccine is recycled from a reservoir beneath the belt to ensure an effective and economic utilization. Protection obtained in this way is found to be comparable to direct immersion.[8]

A second method automates the dipping process itself. A lightweight unit is either floated into a holding tank or set at the edge in such a way that fish are loaded onto a rubber belt that lifts them, through the vaccine bath and thence to a separate tank or partitioned area. Both techniques could advantageously be combined with grading equipment.

TEMPORAL STRATEGIES FOR VACCINATION

It is important for the fish farmer to plan the best times to vaccinate and to give boosters. With respect to *Vibrio* and ERM vaccines the longevity and effectiveness of single applications using standard protocols are sufficient to meet the needs in rainbow trout culture where fish are marketed within 1 year. In other species e.g. Atlantic salmon, the farming cycle lasts for several years and little information yet exists on duration of protection beyond 1 year. However, in many diseases, continued exposure to the pathogen in the environment is expected to act as a natural booster to immunity though at the moment there is no evidence that this actually occurs in fish. With other, less potent vaccines, booster immunization may be necessary to establish acceptable levels of protection and strategies of administering multiple doses will need to be formulated. Ideally, vaccination should fit into the particular husbandry strategies employed on any particular farm, but two factors which change with time are of major and general importance: the size of the fish and the water temperature.

Size of fish

The immune response of fish takes some time to mature after hatching. Although rainbow trout have been shown to become immunocompetent at as little as 0.13 g the level of protection achieved is lower and the duration shorter than in older, larger individuals. Size, not age, is the significant factor. The most complete work, by Johnson *et al.*[9] using six different salmonid species shows that a high level of immunity is achieved only in fish weighing more than 1 g. However, the immunity induced at this size appears to decline after approximately 3 months, whereas fish vaccinated at 4 g retain immunity at a higher level for at least a year (see also Ch. 2).

These factors can easily be translated to farm practice. Immersion vaccination is effective at small sizes. There are three reasons why farmers desire to vaccinate young fish: (a) to protect against diseases affecting young fish; (b) to vaccinate a large number of fish/litre of vaccine; (c) it is logistically easier to mass vaccinate small fish. However, if immunity is required to last for most of a growth season then consideration must be given to the possibility that salmonids less than about 2.5 g may lose protection too soon. Furthermore, since this is a minimum weight, the natural size variation present, even in a freshly graded population, requires that the mean individual weight needs to be higher to ensure that all fish are above the minimum. (More precise values on a particular farm can readily be obtained by weighing a few individuals, estimating variance and calculating the 'z' score.) It is usually recommended therefore that trout fingerlings should have a mean, individual weight of not less than 3.5 g. In practice many hatcheries prefer to sell-on fingerlings at 4–5 g and vaccination is carried out on freshly graded fish of this size, 4 weeks before transfer to a potentially infected site.

Despite these clear recommendations a strategy of vaccinating smaller fish is sometimes practised with ERM vaccination in trout where the on-growing site is persistently infected. Here fish are vaccinated at the hatchery at as little as 1–1.5 g no more than 8 weeks before transfer. The level of immunity acquired, sometimes combined with antibiotic treatment at transfer, protects the fish initially and reliance is placed on a natural, background challenge to boost the immunity continually throughout the season. Evidence that this actually occurs, however, is scanty and non-vaccinated survivors of an outbreak are not usually immune to further attacks.

The individual husbandry requirements of different species frequently dictate the vaccination strategy within these general constraints. Atlantic salmon, although susceptible to ERM, are most frequently vaccinated only against vibriosis. Since the maximum risk of mortalities from vibriosis occurs on transfer to sea water vaccination is timed 4–6 weeks before this event, if possible just in advance of smolting. At this time the fish will vary between 25–40 g. Due to the individual value of the fish, bath vaccination is still considered economic even where only 45 kg fish per litre of vaccine are immunized.

Marine fish, such as turbot, sea bass and sea bream, create different problems. Some hatcheries operating on recycled, sterilized water are effectively disease free, so that as in trout, vaccination is timed according to transfer from the hatchery. Other, particularly turbot, farms are frequently exposed to vibriosis throughout the life cycle. In these species husbandry methods are evolving and vaccination procedures with them. It has been clear since the early development of *Vibrio* vaccines that efficacy is not affected by administration in fresh or sea water but minimum sizes, optimum sizes, the planned use of boosters and longevity of protection are presently areas of incomplete knowledge in species other than salmonids.

Temperature

Most aspects of the physiology of fish, including the rate of acquisition of immunity are affected by temperature. In general, protective immunity develops more slowly as the temperature falls below the optimum for any particular species (see also Ch. 3) and practical interest has centred on this since many potentially vaccinable diseases become significant at rising temperatures and it is clearly advantageous to establish protection in advance of this. Also, the husbandry patterns of salmon and trout, particularly in higher latitudes frequently dictate that vaccination should be carried out, if at all, during the winter–spring months. Amend and Johnson[10] showed that sockeye salmon (*Oncorhynchus nerka*), vaccinated with *Vibrio* bacterin at only 4°C developed immunity after about 40 days. Horne *et al.*[11] also showed protection against vibriosis in trout vaccinated at 4–6°C and in this case protection was maintained for at least 330 days. It is possible therefore to vaccinate salmonids at low temperatures to accommodate management requirements and in advance of an expected disease risk providing it is recognized that a much longer time is required to build up a significant level of immunity. There appears to be no 'cut-off' point in salmonids below which immunity will not be induced, the response merely becomes progressively poorer. However, most manufacturers indicate a point, usually 5–6°C, below which vaccination is not recommended and, where the option exists, as high a temperature as possible, up to the optimum for the species, should be used.

ECONOMIC STRATEGIES OF FISH VACCINATION

Whether or not a fish farmer will use a commercial vaccine is basically an economic decision and a cost : benefit analysis has to be carried out. The relevant factors in such an analysis will vary from farm to farm and only a broad perspective can be given here. No vaccines for animal or human are 100 per cent effective. Some individuals in any population will not be effectively protected for prolonged periods and some disease will still occur making other forms of disease control e.g. antibiotic therapy, necessary.

However, a cost effective vaccine will not only reduce economic losses through mortalities, but should also alleviate decreased growth rates due to infection, and reduce the need for chemotherapy. Thus, the cost : benefit analysis needs to be based upon certain expectations of how well a vaccine might perform given certain conditions under which the farmer is considering its use. Such expectations need to be based upon the results of large field trials which are appropriate to the type of farm in question. To date most available data comes from the use of ERM vaccines in rainbow trout farms. A detailed economic analysis of this data has been made by Horne and Robertson[12] for ERM.

For the basis of calculation it is assumed that the cost of ERM and vibriosis vaccines is about £100/litre. The cost per fish depends upon the method of delivery and the size of the fish (Table 5.3). The injectable vaccine dose is 0.1 ml/fish irrespective of the fish's size (i.e. 1 p/fish), while the dose of the immersion vaccine is calculated at 1 litre/100 kg fish i.e. 1 p for a 10 g fish, 10 p for a 100 g fish. Thus the most cost-effective means for vaccinating fish below 10 g is the immersion method, whereas for fish progressively larger than 10 g the injection route becomes more and more cost effective, though in practice it is not realistic to inject fish under 20 g.

Table 5.3 *Cost and doses of ERM immersion vaccination for fish of different sizes*

Size of fish		Vaccine	
No./lb	g	Vol (litre)/1000 fish	£/1000*
130	3.5	0.035	3.5
100	4.5	0.045	2.5
90	5.0	0.050	5.0
60	7.5	0.075	7.5
40	11.0	0.04	11.0
30	15.0	0.150	15
20	22.0	0.220	22
7.5	60.0	0.600	60
4.5	100.0	1.000	100
1.8	250.0	2.500	250
1.0	450.0	4.500	450

*Vaccine is costed at £100/litre. Eleven vaccinates: 100 kg fish by immersion, 10 000 fish by i.p. injection.

In an economic analysis of immersion vaccination of rainbow trout of 4.5 g against ERM (fish being reared to a mean individual weight of 250 g) based upon available field data Horne and Robertson[12] calculated that if vaccinating fish resulted in 1 per cent more fish reaching market the cost of the vaccine would be recovered.

Moreover, increased survival of vaccinated fish is not the only economic benefit. The field data on ERM vaccinated rainbow trout indicates that savings in the cost of medicated feed of up to 70 per cent may be expected and also increases in food conversion ratios.

CONCLUSIONS

Throughout this discussion it has been implied that, providing fish are above a certain minimum size when vaccinated, protection is long lasting in relation to the fish management cycle. However, duration of protection has not been extensively investigated over a wide range of species and conditions. Since protection declines gradually it should be borne in mind that properly vaccinated fish, held for some months unexposed to the target disease, may succumb in large numbers if presented with a sudden severe challenge such as may occur if transferred late to an infected site or exposed to a batch of imported, infected fish.

ERM and *Vibrio* vaccines have been proven to be highly cost effective in areas where losses from these diseases might be expected to be high and where fish at the time of vaccination are healthy. The precise values for a cost : benefit analysis will depend upon the performance of the vaccine and the food costs, medication costs and the stock value. When comparing alternatives, antibiotics are generally used only after disease has become apparent, may require several applications and become progressively more expensive as fish size increases. Vaccination requires only a single initial treatment but, since it is not 100 per cent effective, some antibiotic treatment may still be required. However, the benefits from vaccination of reduced morbidity and mortality resulting in improved growth performance, reduced use of medicated food and better utilization of holding facilities frees time and resources on a farm to be devoted to overall improvements in husbandry which ultimately is the most effective form of disease control.

REFERENCES

1 Zapata, A. G., Torroba, M., Alvarez, F., Anderson, D. P., Dixon, O. W. and Wisniewski, M. (1987). Electron microscopic examination of antigen uptake by salmonid gill cells after bath immunisation with a bacterin. *J. Fish. Biol.*, **31** Suppl. A, 209–217.

2 Tatner, M. F. (1978). Quantitative relationship between vaccine dilution, length of immersion time and antigen uptake using a radiolabelled *Aeromonas salmonicida* bath in direct immersion experiments with rainbow trout, *Salmo gairdneri* *Aquaculture,* **62**, 173–185.

3 Tytler, P. and Blaxter, J. H. S. (1988). Drinking in yolk sac larvae of halibut, *Hippoglossus hippoglossus*. *J. Fish Biol.,* **32**, 493–494
4 Egidius, E. C. and Andersen, K. (1979). Bath immunisation – a practical and non-stressing method of vaccinating sea farmed rainbow trout (*Salmo gairdneri* Richardson) against vibriosis. *J. Fish Dis.,* **2**, 406–410.
5 Duff, D. C. B. (1942). The oral immunization of trout against *Bacterium salmonicida*. *J. Immunol.* **44**, 87–94.
6 Agius, C., Horne, M. T. and Ward, P. D. (1983). Immunisation of rainbow trout, *Salmo gairdneri* Richardson, against vibriosis: comparison of an extract antigen with whole cell bacterins by oral and intraperitoneal routes. *J. Fish Dis.,* **6**, 129–134.
7 Gould, R. W., O'Leary, P. J., Garrison, R. L., Rohovec, J. S. and Fryer, J. L. (1978). Spray vaccination: a method for immunisation of fish. *Fish Path.*, **13**, 62–68.
8 Tebbit, G. L. and Goodrich, T. D. (1983). Vibriosis and the development of effective bacterins for its control. In: *Antigens of Fish Pathogens*, Collection Fondation Marcel Merieux, pp. 225.
9 Johnson, K. A., Flynn, J. K. and Amend, D. F. (1982). Onset of immunity in salmonid fry vaccinated by direct immersion in *Vibrio anguillarum* and *Yersinia ruckeri* bacterins. *J. Fish Dis.,* **5**, 197–205.
10 Amend, D. F. and Johnson, K. A. (1981). Current status and future needs of *Vibrio anguillarum* bacterins. *Dev. Biol. Standard,* **49**, 403–417.
11 Horne, M. T., Tatner, M., McDerment, S., Agius, C. and Ward, P. D. (1982). Vaccination of rainbow trout, *Salmo gairdneri* Richardson, at low temperatures and the long persistence of protection. *J. Fish Dis.,* **5**, 343–345.
12 Horne, M. T. and Robertson, D. A. (1987). Economics of vaccination against enteric redmouth disease of salmonids. *Aquacult. Fish. Manage.,* **18**, 131–137.

6

Vaccination against Vibriosis

P. D. Smith

THE DISEASE PROBLEM

Vibriosis is a bacterial disease of salt-water and migratory fish and the severity of vibriosis has increased proportionately with the development and expansion of fish farming worldwide. In severe cases of vibriosis there may be up to 40 per cent mortality. Recent figures suggest that in Japan alone annual losses due to the disease exceed £11 million.

History

The first reference to a bacterial disease of fish was a description of vibriosis of 'Red Pest' by Bonaveri.[1] 'Red Pest' caused severe mortalities among eels in seawater ponds in Italy during the 18th and 19th centuries. The causative agent of 'Red Pest' in eels was isolated by Canestrini[2] and designated *Bacterium anguillarum*. The bacterium was subsequently isolated from an outbreak of the disease in eels in Sweden by Bergman[3] and re-named *Vibrio anguillarum*.

Geographical and host range

Vibriosis has a global distribution with epizootics on all continents and in a wide range of fish species. An excellent review of the distribution of vibriosis in fish is provided by Anderson and Conroy.[4]

The host range of vibriosis includes both cultured and feral fish. The disease causes significant losses in cultured Pacific salmon, Atlantic salmon and rainbow trout grown in marine cage farms and in cultured non-salmonids including eel, yellowtail, red sea bream, and sea bass.[5–7]

There are several reports of vibriosis in a considerable range of feral fish.[4,5,8–16] Reports of vibriosis have not been limited to marine environments. Rucker[17] and Ross *et al.*[18] reported epizootics of vibriosis in rainbow trout raised in fresh water possibly due to the feeding of contaminated marine trash fish. Hacking and Budd[19] described an outbreak of vibriosis in tropical fishes maintained in a freshwater aquarium. In addition, problems of vibriosis caused by *Vibrio anguillarum* have occurred in freshwater farmed

FISH VACCINATION
ISBN 0-12-237485-1

rainbow trout in Italy.[20,21] Vibriosis is also a significant problem in farmed crustaceans e.g. prawns and shrimp.

Vibrio species/strain variation

The most commonly encountered fish pathogenic *Vibrio* species is *V. anguillarum* and it is this species which is responsible for the majority of losses worldwide, and was probably the causative agent of the disease featured in the original reports of 'Red Pest' of eels in the 18th and 19th centuries.

V. anguillarum constitutes part of the normal microflora of the aquatic environment with maximum and minimum numbers in summer and winter respectively.[22] *V. anguillarum* may constitute part of the natural microflora of marine fish[23,24] and consequentially, fish may be continually threatened by the pathogen which may be able to cause disease under certain conditions.

The next most important member of the *Vibrio* species is *V. ordalii* (formerly known as *V. anguillarum* Biotype II) which has been documented in Japan and the Pacific northwest, USA. Its occurrence is not as frequent as that of *V. anguillarum*. Other *Vibrio* species reported to be pathogenic to fish are as follows:

V. alginolyticus which has been implicated in vibriosis of sea bream (*Sparus aurata*) in Israel[25] and of sea mullet.[26]
V. carchariae isolated from diseased sharks.[27]
V. cholerae (NON-01) isolated from diseased ayu in Japan.[28]
V. damsela isolated from damselfish in southern California.
V. vulnificus Biogroup 2 isolated from eels in Japan[28-30] and the United Kingdom (B. Austin, unpublished data).
V. salmonicida recently characterized and named by Egidius *et al.*,[31] is the causal agent of 'Hitra' disease or coldwater vibriosis principally in Norwegian farmed salmon.

Pathogenicity

Symptoms of vibriosis are similar to those caused by *Aeromonas salmonicida*. The external pathology produced by *V. anguillarum* includes haemorrhaging at the base of the fins, around the vent and gills and inside the mouth. Petechiae, necrotic lesions and diffuse haemorrhages can appear on the body surfaces. Internally, the intestine is often inflamed with petechiae present on the viscera and musculature. The intestine may be distended and filled with clear viscous fluid. This histopathology produced by *V. anguillarum* and *V. ordalii* in salmonids is somewhat different.[32]

Experimentally induced infection by water-borne exposure demonstrated that both species entered the fish by penetrating the descending intestine and rectum while *V. ordalii* could also enter by penetrating the skin.[33]

Epidemiology

Vibriosis is a stress-related disease and is associated with a number of factors e.g. high water temperatures, rapid changes in water temperature or salinity, over-crowding, buffeting by wave action and poor water quality including low oxygen levels and high suspended solids.

There also appears to be a seasonal effect on the disease, outbreaks occurring mainly in spring and autumn, though this may simply reflect rapid changes of temperature and salinity. Salmon smolts are particularly sensitive to vibriosis possibly due to some form of immuno-suppression caused by the rapid physiological changes taking place in these fish.

There also appears to be a variation in susceptibility to the disease between salmonid species e.g. Atlantic salmon appear slightly more susceptible than rainbow trout. There does not appear to be an association between age of the fish and susceptibility or severity of the disease.

The type of holding facility may affect the severity of the disease. For example, net damage may act as a focus of infection. In addition, there is a possibility that *Vibrio* bacteria can survive for a considerable time in the slime of uncleaned tanks and fouled nets, thus acting as a reservoir of infection. In particularly sheltered areas where there is little tidal action to flush the fish farm site there can be a considerable build-up of unused food and faeces under the floating cages. There is some suggestion that marine vibrios may grow in such deposits, which once again may act as reservoirs of infection.

Other sources of infection are carrier and feral fish. If carrier fish are stressed or weakened in any way the carrier state may develop into clinical disease with the infection spreading to other fish through the water. Reports of vibriosis in feral fish and the close proximity of escapees round a marine cage site mean that a farm could always be exposed to infection from external sources.

Chemotherapy

Prior to the introduction of immunization, the most common method employed to control vibriosis was treatment of diseased fish with antibiotics and antimicrobial chemicals.

Antibiotics and antimicrobials are commonly administered as food additives or in bath treatments. Commonly used substances are tetracycline, sulphonamides, nitrofuran derivatives, trimethoprim and, more recently the quinolines (e.g. oxolinic acid). However, antibiotic therapy for the control of vibriosis has met with variable success.

The most significant difficulty caused by the continued and indiscriminate use of antibiotics has been the development of serious drug resistance problems with *V. anguillarum*. Aoki *et al.*[34] reported that 65 of 68 random *V. anguillarum* isolates from vibriosis epizootics in Japan carried transferable drug resistance factors (R plasmids). In a further study[35] 259 isolates of *V. anguillarum* obtained from diseased ayu in Japan from 1974 to 1977 were tested for their sensitivity to 12 chemotherapeutic drugs. Nine strains were sensitive to all of the drugs tested and the remaining 250 isolates were resistant to various combinations of six commonly used compounds i.e. chloramphenicol, tetracycline, nalidixic acid, furazolidone, sulphamonomethoxine and trimethoprim. Transferable R plasmids were found in 165 of 250 strains and the most common type of plasmid-determined resistance was to chloramphenicol, tetracycline and sulphamonomethoxine.

With the increasing expansion of commercial aquaculture for the production of food fish there was an urgent requirement for more effective methods for the control of vibriosis. In many cases, control of the disease by management practices has not proven practical and problems presented by drug resistance have resulted in attempts to control vibriosis by immunization and the development of commercial vaccines.

NATURE OF VIBRIO VACCINES

Current commercial *Vibrio* vaccines in the Northern hemisphere contain mixtures of the most commonly encountered species which cause disease problems i.e. *V. anguillarum* and *V. ordalii*. The majority of these vaccines are simple inactivated cultures containing mixtures of whole cells and extracellular products. In a few cases the bacteria have been lysed either by chemical or physical methods.

The protective antigens of *V. anguillarum* appear to be heat-stable (to 100–121°C) lipopolysaccharides in the cell wall, which may also be released into the culture supernatant.[36,37] These large molecular-weight compounds i.e. 100 kilodaltons (kd)[38] are able to withstand severe extraction methods. Chart and Trust[36] isolated, from the outer membrane, two other minor proteins with molecular weights of 49–51 kd which were potent antigens. A weakly antigenic protein with a molecular weight of approximately 40 kd was also present. These antigens may be heat labile, which could account for the greater protection achieved with formalin-inactivated vaccines compared to heat-killed products.[39,40]

Development of live attenuated vaccines is largely academic as there are likely to be problems with regulatory authorities regarding licensing for fisheries' use.

It is unlikely that *Vibrio* vaccines will become more sophisticated than the present simple formulations. While the use of genetic engineering techniques to produce antigens in bulk has been suggested, it is doubtful whether a vaccine could be produced by such methods at a price which would be acceptable by the aquaculture industry.

An area where we may expect an improvement in formulation is the addition of water-soluble adjuvants to immersion-applied vaccines in order to improve potency and duration of protection.

While some experimental work on vaccines containing *Vibrio* species other than *V. anguillarum* and *V. ordalii* has been carried out there are as yet no commercially available vaccines containing other *Vibrio* species known to be pathogenic to fish, although work in Norway is advanced in the production of a vaccine against coldwater vibriosis caused by *V. salmonicida*, with successful field trials reported using a formalized whole culture bacterin delivered by immersion.[41]

VACCINE DEVELOPMENT

The evolution of commercial *V. anguillarum* vaccines commenced with the development and utilization of mass immunization methods, necessary to have practical application in modern aquaculture systems where large numbers of fish are involved.

Several methods of administering commercial *Vibrio* vaccines are currently employed. These are: injection, oral, hyperosmotic immersion, immersion or dip, bath vaccination, spray or shower and automated immersion.

Injection

Intraperitoneal (i.p.) injection was used in the first reported effective immunization of fish against vibriosis.[42] Disadvantages of injection include size limitation (it is only feasible to inject fish of 15 g and larger), labour intensity, and excessive stress caused by handling and anaesthesia. Handling is particularly important in the case of Atlantic salmon which are particularly sensitive to scale damage and infection at the smolting stage.

However, injection vaccination ensures that each fish receives a constant and exact dose of product and when the efficacy of vaccination methods is compared, injection vaccination proves to be the most effective.

Many of the disadvantages of injection vaccination can be overcome. For example, vaccination may be timed with grading, a time when the fish are normally being handled. The process can also be speeded-up and made simpler for the operator by the use of multi-dose repeater syringes identical to those used in the poultry industry.

The key to successful injection vaccination rests in the development of a 'system' which will vary according to the size and nature of the fish farm site (see Ch. 1). Despite the practical limitations of injection, the method has been used in the Pacific northwest of the USA to vaccinate Pacific salmon against vibriosis since the early 1970s and, more recently, the method has become common practice in Scandinavia.

A further advantage of injection vaccination is that injectable vaccines permit the use of certain adjuvants that cannot be used with other delivery systems. Indeed an injectable *Vibrio* vaccine in combination with *Aeromonas salmonicida* is now commercially available and this vaccine contains an oily adjuvant to improve efficacy and duration of immunity.

However, some of the disadvantages of injection vaccination plus an increasing requirement for vaccinating small fry at the hatchery led to the development of mass administered vaccines.

Oral vaccines

Oral vaccination was the first mass immunization method employed to vaccinate against vibriosis.[43–46] This work demonstrated that both formalin-killed lyophilized whole cells and wet packed whole cells of *V. anguillarum* provided some degree of protection.

Maximal protection of chinook salmon followed the feeding of 2 mg of dried vaccine/g of food for 15 days at temperatures even as low as 3.9°C.[46] An important finding was the observation that longer feeding regimes did not result in enhanced protection.

However, when the immune response of parenterally immunized and orally immunized fish was compared, the former had high serum titres of agglutinating antibody while antibody was undetectable in the latter. These results indicated that different immune responses were produced by the two immunization methods.

Oral vaccination has important theoretical advantages over other methods of vaccination, principally in requiring no stressful handling and no break in routine. However, most trials have resulted in low efficacy of oral vaccination (see below) which may result from three important limitations; poor potency, lack of controlled dosage and the stability of the antigen. The poor potency and problems associated with different feeding rates of different fish within a population can only partially be overcome by long term treatment and this also consumes large quantities of relatively expensive vaccine.

Possibly the way ahead for oral vaccines rests with the food manufacturers. A method must be developed which allows incorporation of the bacterial antigens into food without harsh processes such as heating which may destroy antigenicity. Palatability is also a problem which has been encountered with previous experimental oral vaccines.

Another problem which may exist with oral vaccines resides in the location of the immune areas of the fish gut. It appears these are located in the lower intestine, and thus to reach them and therefore induce an immune response, the antigens must first pass through the stomach of the fish. It is possible that the high acid conditions in the stomach may inactivate antigens. This hypothesis is supported by the increased efficacy of *Vibrio* vaccines administered by anal intubation rather than orally.[47] A way of overcoming the inactivation of vaccines in the stomach may be microencapsulation to protect their passage through the stomach and to release the antigens in the immune areas of the lower gut. However, as yet, work on microencapsulated vaccines has been purely experimental and there is little prospect of a commercially-available product becoming available in the near future.

It is possible that the previously mentioned disadvantages of oral vaccines may preclude their use as a primary vaccination method. However, they may well have a place in fish health management when used as a booster on fish previously vaccinated by immersion (see below). To expand this point further, many species of fish (e.g. salmon) are required to grow for long periods before reaching table size and this period often exceeds the 12 months protection afforded by immersion vaccination. There is, therefore, a requirement to give a booster vaccination at approximately 1 year after the primary vaccination. At this time it may not be practical or wise to handle fish for booster vaccination by either injection or direct immersion. Therefore, oral vaccination may be a useful method of booster vaccinating such fish.

Whatever the case, the demand for orally administered vaccines is sufficient to warrant continued research. It is probable that the industry will see commercially available oral *Vibrio* vaccines in the near future.

Hyperosomotic–immersion vaccines

This method is purely of historical interest as it has been superceded by methods such as simple immersion or bath vaccination.

The technique involves the immersion of fish in a hypertonic saline solution for a short time followed by a dip in an aqueous solution of the vaccine. The rationale behind the technique was that in the hypertonic salt solution fish lose water by osmosis, and when placed in the vaccine solution the fish replace water lost by osmosis at the same time taking up vaccine.

While the hyperosmotic vaccination method gave good and reasonably long-lasting protection against disease, there were problems including hypertonic stress and gill damage which could reduce the growth rate of fish or make them susceptible to other infections. Because of these problems, research continued into modification of the hyperosmotic vaccination method and gave rise to the immersion or dip vaccination methods commonly in use today.

Immersion or dip vaccination

It subsequently became clear that the stressful dip in the hypertonic saline was unnecessary and equivalent protection rates could be obtained when the fish were dipped directly in a solution of vaccine.

The technique simply involves dilution of the vaccine with hatchery water (usually 1:10 for most of the commercially-available vaccines), removal of the fish from the holding facility, immersion of the fish in the vaccine solution for up to 30 s, draining of the excess vaccine from the fish and the return of the fish to the holding facility. The main portal of antigen uptake in dip vaccination is the gill tissue.[48,49] Work with radio-labelled antigens suggests an active uptake mechanism by the gills possibly involving phagocytic cells on the gill surface. Immediately after vaccination, labelled antigens are concentrated in cells on the gill surface. A short while after vaccination these cells appear in the blood capillaries of the gill and some hours later in the anterior kidney and spleen, presumably where they are available for immune processing.

Attempts have been made to improve the uptake of the vaccine by the addition of surfactants presumed to act at the gill surface but it is debatable whether this has any marked effect on protection levels.

The immersion method of vaccination can be scaled up or down or fully automated depending on the size and nature of the fish farm site where the fish are to be vaccinated.

On a small scale, vaccination can be performed using hand nets and small tanks to contain the diluted vaccine. This can be scaled up to utilize large containers or dustbins and a custom-made net liner that fits the container. For large sites, automatic vaccination machines (Wildlife Vaccines Inc.) can be used (Fig. 6.1). Such machines consist of a reservoir containing diluted vaccine through which passes a conveyor belt driven by an electric motor. Fish are netted onto the conveyor belt which then passes through the vaccine to give a set exposure time which is controlled by the speed of the electric motor. Such automated systems or modifications of them may well represent the way ahead for immersion applied vaccines as standard net dipping techniques, when performed on a large scale, can be quite labour intensive. There are some problems with an automated system, notably, of transmission of disease from farm to farm and therefore the requirement to thoroughly sterilize each machine after use.

Dip vaccines have been designed such that fish can be successfully vaccinated with a relatively short exposure e.g. 30 s to a relatively concentrated solution of vaccine e.g. a one in 10 dilution. Extension of this time in concentrated vaccine will not significantly improve protection or duration of protection. However, a much more dilute vaccine solution can be used provided that the time of exposure is extended and this is the basis for the bath method of vaccination.

Fig. 6.1 Automatic immersion vaccination machine.

Bath vaccination

Bath vaccination was developed because of the problems associated with net damage and handling stress when vaccinating salmon smolts by dip method. It is advisable to vaccinate salmon smolts against vibriosis while they are in fresh water, just prior to transfer to sea water. Unfortunately, smolts are particularly sensitive to scale damage from netting at this time and handling of this nature should be avoided if at all possible.

Bath vaccination allows vaccination of fish within their holding tanks without removal. Basically, the technique requires reducing the level of the water in the holding tank to a known volume and adding vaccine to a known concentration (usually between 1:200 and 1:1000 depending on the commercial product used). The fish are allowed to swim in this highly diluted vaccine for 1 to 2 h after which water flow is returned to the tank and the volume restored to normal.

While this technique requires little effort on the part of the fish farmer and is relatively easy to perform, there are still some doubts as to the efficacy and duration of immunity afforded by this method (see below). In addition, some care needs to be taken while performing bath vaccination. This method requires a large number of fish to be held in a relatively small volume of water for some considerable time and careful attention should be paid to the oxygen levels in the water. Ideally the water should be constantly oxygenated during vaccination, and if possible, the oxygen levels should be monitored during the whole procedure with a view to immediate restoration of the water supply should fish show signs of stress.

Bath vaccination may have a number of applications in addition to being a safe method for the vaccination of salmon smolts. The method appears promising for the vaccination of larval fish or crustaceans where netting would pose a problem. The technique may lend itself to booster vaccinating salmon in cages, by a procedure similar to that used for administering salmon louse treatments.[50]

Finally, an obvious application of bath vaccination is the vaccination of fish in road (or sea) transport tanks. However, it should be borne in mind that the induction of the immune response in fish takes some time depending on the ambient water temperature and therefore fish vaccinated in transport tanks arriving at a site where an immediate epizootic is expected will not be immune to the disease.

Spray vaccination

Spray vaccination was a development of a method used for marking fish with a dye and was developed simultaneously with immersion vaccination. This method involves the spraying of vaccine onto fish skin and can be semi-automated using a conveyor belt which passes under spray nozzles.

Initially high pressure spraying was developed[40,51] but low pressure spraying is equally effective in conferring protection against vibriosis in coho salmon, ayu, and rainbow trout.[51,52] However, spray vaccination can be stressful to fish and results obtained using this method can be variable.

VACCINE TRIALS

Immunity

In the case of vibriosis a protective role of antibody has been clearly demonstrated. Harrell *et al.*[53] passively immunized rainbow trout with immune serum and observed protection upon challenge. The protection was specific as the effect could be abrogated by preincubation of the immune serum with *Vibrio* bacterin. Moreover, the decay of the injected immunoglobin correlated with the decrease in protection. Similar results using antibody were reported by Viele *et al.*[54] In addition, transfer of lymphoid cells from the head kidney of immunized fish also conferred protection upon non-immunized donors although cells from the spleen and thymus did not. The above experiments show that the humoral response provides good protection against vibriosis. However, this correlation is not always clear. Immersion vaccination can be followed by serum antibody production[55] but there are cases in which protection was observed in the absence of serum antibodies.[51,56]

In most studies on oral vaccination against vibriosis, no serum agglutinating antibody was observed although a certain degree of protection was obtained.[39,46,51,57] Fletcher and White[58] demonstrated specific antibody in intestinal mucus after feeding of *Vibrio* bacterin while serum antibody levels were low, suggesting a local production of antibody. Kawai *et al.*[57] found that orally vaccinated ayu developed an anti-*Vibrio* agglutinin in skin mucus, whereas no antibody was found in serum or intestinal mucus, although these fish were protected. Serum from orally vaccinated ayu, which did not contain anti-*Vibrio* agglutinins, was transferred to non-immunized fish and conferred protection after challenge. Moreover, in vaccinated fish the colonization of the skin by viable bacterial cells was significantly reduced, especially in the mouth region and on the gills. Also in the stomach and intestine no bacterial colonization occurred, despite the fact that in intestinal mucus no anti-*Vibrio* activity could be detected.[57] At the moment these results are difficult to explain but improved techniques for assaying specific antibodies in low concentration in mucus secretions may yet confirm the protective role of antibody.

Effectiveness

As previously mentioned there are a number of vaccination methods available for use on a commercial scale. Some published trials have directly compared these methods and are useful for evaluating which method would be of most use on a commercial scale and for calculation of the economic benefits of vaccination.

First, it must be pointed out that on a commercial scale no fish vaccine is ever 100 per cent effective and within a vaccinated population exposed to high numbers of pathogens, one should always expect to see some fish succumbing to the disease. The extent of this will depend on a number of factors which will be discussed in a later section. Evidence from trials shows that oral vaccination is the least successful while injection is the most effective (Table 6.1).

Table 6.1 *Comparison of effectiveness of different* Vibrio *vaccine trials using different methods of delivery*

	Mortality (%)			
	1[59]	2[60]	3[61]	4[62]
Unvaccinated controls	33.8	52	100	54.7
Oral	31.7	27	94	—
Immersion	2.1	4	53	5.2
Bath	—	—	—	15.2
Spray	—	1	—	—
Injection	1.4	0	7	4.3

A comparison of the bath method of vaccination with immersion vaccination and injection by Hastein and Refstie[62] (Table 6.1) shows the bath administration was the least effective although the level of protection may be acceptable in commercial operations when offset against the lack of stress and handling involved.

Work on the duration of immunity has been scanty and most of this information is being obtained from commercial use of the vaccines. In a study by Johnson *et al.*,[63,64] the duration of protective immunity varied with the vaccine concentration and the size and species of fish used. They found that when fish were vaccinated at 4 g and larger, immunity lasted for a year or longer. The duration of immunity probably depends on the level of natural challenge present and this will be discussed in a later section.

While not directly connected with specific immunity and protection one further aspect of vaccination cannot be ignored and this is the demonstration of an improved, so called 'Health Index' in vaccinated fish. This appears to apply for all fish vaccines but in reports of trials of *Vibrio* vaccines, vaccinated fish do appear to be more resistant to other diseases – bacterial, fungal and parasitic – and also have a better food conversion when compared with unvaccinated control fish. There are a number of theories put forward to explain this phenomenon. First, and most likely, is that the major disease problem is being taken away from the fish and these 'healthier' fish are better able to cope with other disease problems. One other theory is that because *Vibrio* spp. are Gram negative organisms (which contain lipopolysaccharide in their cell wall and which is a B lymphocyte mitogen) they may be conferring a non-specific adjuvant effect and thus stimulating the fish's immune response to counteract challenges with other organisms.

LIMITATIONS OF PRESENT VACCINES

There are a number of factors which have a marked influence on the level and duration of protection of existing commercial *Vibrio* vaccines. All these factors should be borne in mind when determining a suitable vaccination schedule.

Size and age

Johnson *et al.*[60] studied the ontogeny of immunity in salmonid fry of six different species, vaccinated by immersion against vibriosis and enteric redmouth disease. The minimal fish size at which protection occurred was between 1.0 and 2.5 g. Here, immunity appeared to be a function of size rather than age. The duration of protective immunity varied with fish size. Generally it lasted for about 120 days in 1 g fish, about 180 days in 2 g fish and longer than a year in 4 g fish.[64]

The efficacy of vaccines when administered to fish of less than 2 g, can be improved by giving the fish a booster vaccination 14 days after the first vaccination (Tebbit, personal communication). However, booster vaccinations double the cost of vaccination and also require further handling and netting which is stressful to fish.

Tatner and Horne[65] found that rainbow trout fry were not susceptible to *V. anguillarum* administered by i.p. injection or by bath until they reached the age of 7 or 8 weeks post-hatch respectively. Vaccination by direct immersion of 2, 4, 6 and 10 weeks old fry resulted in increasing protective immunity upon an i.p. challenge. I.p. vaccination of 10 weeks old fry resulted in 100 per cent protection.

The size related factor is a significant problem for species of marine fish which spend all of their life cycle in salt water. In theory, unless hatchery water is sterilized, these fish are under constant threat from vibriosis and therefore they should be vaccinated as soon as possible after hatching. More work is required to study the onset of immunity in such fish to allow an effective vaccination schedule to be developed.

Adjuvants

Commercially available injectable vaccines contain various adjuvants depending upon the manufacturer of the vaccine. There is no doubt that the inclusion of the adjuvant increases the level and the duration of the efficacy.

Presently, adjuvants are available only for injectable vaccines. There are a number of water soluble adjuvants which could conceivably be added to immersion or oral vaccines to improve their protection and duration of protection and this is likely to be an important area of future research.

Temperature

As fish are poikilothermic, temperature has a profound effect on the length of the induction period of immunity and therefore the planning of vaccination programmes. In general the colder the water temperature, the longer the time for the development of protective immunity. Thus, in salmonids at water temperatures of 10°C it will take 14 to 21 days for the development of protective immunity to *Vibrio* vaccine. If the temperature is lowered to 5°C this time will be extended to as much as 40 days, while higher temperatures will shorten the induction period.

Obviously, this phenomenon must be taken into consideration when vaccination programmes are being planned. Fish are most often vaccinated on disease free hatcheries, prior to their transport to growing-on sites where they may encounter disease epizootics. Vaccination must therefore be planned to allow sufficient time for protective immunity to develop before transfer to an infected site. Obviously, at low water temperatures this may take many weeks.

At very low temperatures, 4°C and less, the immune system will not react rendering vaccination ineffective. This may represent a significant constraint on the successful use of *Vibrio* vaccines particularly in northern parts of the Northern hemisphere e.g. Canada, Alaska and Scandinavia where water temperatures can be particularly low. A solution to the problem would be to raise water temperatures for a period during vaccination, but unless cheap sources of low grade energy are available (e.g. geothermal springs or industrial waste heat) this would be prohibitively expensive.

CONCLUSIONS

Vibrio vaccines are currently the most successful and show the most promise of any vaccine against the major diseases of fish. Commercial vaccines are well-established and well-researched and data from extensive commercial use is constantly being used to improve administration and administration methods. *Vibrio* vaccines have the double advantage of being simple to manufacture and simple to use in a variety of mass administration methods.

With the enormous expansion of the marine farming of trout and salmon in the last few years, vaccination has proven itself to be a successful and highly cost-effective tool in the management of the disease.

The future for *Vibrio* vaccines also looks promising. Marine aquaculture is likely to expand rapidly and developments in technology are allowing more exposed areas of sea to be farmed or for sea water to be inexpensively pumped onto land-based marine farms. More and more marine species are being farmed worldwide and the constant presence of *Vibrio* bacteria in seawater makes the disease a constant threat. Moreover, the likelihood of legislation against the indiscriminate use of antibiotics in the aquatic environment will further encourage the use of vaccines.

It is likely we will soon see multivalent *Vibrio* vaccines containing a mixture of *Vibrio* species and a further application of *Vibrio* vaccines is likely to be in the control of vibriosis in invertebrates with the vaccination of cultured shrimps and prawns.

REFERENCES

1 Bonaveri, G. F. (1761). quoted by Drouin de Bouville (1907).
2 Canestrini, G. (1893). La malattia dominate delle anguille. *Atti Institute Veneto Service,* **7**, 809–814.
3 Bergman, A. M. (1909). Die rote Beulen-krankheit des Aals. *Bericht aus der Königlichen Bayerischen Versuchsstation,* **2**, 10–54.
4 Anderson, J. W. and Conroy, D. A. (1970). Vibrio diseases in fishes. In: *A Symposium on Diseases of Fishes and Shellfishes,* ed. S. F. Snieszko, Washington, DC: American Fisheries Society, Special Publication No. 5, 226–272.
5 McCarthy, D. H. (1976). Vibrio disease in eels. *J. Fish. Biol.,* **8**, 317–320.
6 Ezura, Y., Tajima, K., Yoshimizu, M. and Kimura, T. (1980). Studies on the taxonomy and serology of causative organisms of fish vibriosis. *Fish Path.,* **14**, 167–179.
7 Brisinello, W., Doimi, M., Giorgetti, G. and Sarti, M. (1985). Vaccination trials against vibriosis in sea bass (*Dicentrarchus labrax*) fry. *Bull. Eur. Ass. Fish Pathol.,* **5**, 55–56.
8 Cisar, J. O. and Fryer, J. L. (1969). An epizootic of vibriosis in chinook salmon *Bull. Wildl. Dis. Assoc.,* **5**, 73–76.
9 Pacha, R. E. and Kiehn, E. D. (1969). Characterization and relatedness of marine vibrios pathogenic to fish: physiology, serology and epidemiology. *J. Bact.,* **100**, 1242–1247.
10 Evelyn, T. P. T. (1971). First records of vibriosis in Pacific salmon cultured in Canada, and taxonomic studies of the response bacterium, *Vibrio anguillarum. J. Fish. Res. Board Can.,* **28**, 517–525.
11 Håstein, T., and Holt G. (1972). The occurrence of vibrio disease in wild Norwegian fish. *J. Fish Biol.* **4**, 33–37.
12 Levine, M. A., Wolke, R. E. and Cabelli, U. J. (1972). *Vibrio anguillarum* as a cause of disease in winter flounder (*Pseudopleuronectes americanus*). *Can. J. Microbiol.,* **18**, 1585–1892.

13 Harrell, L. W., Etlinger, H. M. and Hodgins, H. O. (1976). Humoral factors important in resistance of salmonid fish to bacterial disease. II. Anti-*Vibrio anguillarum* activity in mucus and observations on complement. *Aquaculture*, **7**, 363–370.
14 Egidius, E. and Andersen, K. (1978). Host-specific pathogenicity of strains of *Vibrio anguillarum* isolated from rainbow trout *Salmo gairdneri* (Richardson) and Saithe *Pollachius viriens* (L). *J. Fish Dis.*, **1**, 45–50.
15 Håstein, T. and Smith, J. E. (1977). A study of *Vibrio anguillarum* from farms and wild fish using principal components analysis. *J. Fish Biol.*, **11**, 69–75.
16 Strout, R. G., Sawyer, E. S. and Coutermarsh, B. A. (1978). Pathogenic vibrios in confinement-reared and feral fishes of the Maine–New Hampshire coast. *J. Fish. Res. Board Can.*, **35**, 403–408.
17 Rucker, R. R. (1959). Vibrio infections among marine and fresh-water fish. *Progr. Fish Cult.*, **21**, 22–25.
18 Ross, A. J., Martin, J. E. and Bressler, V. (1968). *Vibrio anguillarum* from an epizootic in rainbow trout (*Salmo gairdeneri*) in the USA. *Bull. Off. Int. Epiz.*, **69**, 1139–1148.
19 Hacking, M. A. and Budd, J. (1971). Vibrio infection in tropical fish in a freshwater aquarium. *J. Wildl. Dis.*, 7, 273–280.
20 Ghittino, P., Andruetto, S. and Vigliani, E. (1972). "Redmouth" enzootic in hatchery rainbow trout caused by *Vibrio anguillarum*. *Riv. Ital. Piscic. Ittiopathologia*, **7**(2), 41–45.
21 Giorgetti, G., Tomasin, A. B. and Ceschia, G. (1981). First Italian anti-vibriosis vaccination experiments of freshwater farmed rainbow trout. *Develop. Biol. Standard.*, **49**, 455–459.
22 West, P. A. and Lee, J. V. (1982). Ecology of *Vibrio* species, including *Vibrio cholerae*, in natural waters of Kent, England. *J. Appl. Bacteriol.*, **52**, 435–448.
23 Oppenheimer, C. H. (1962). Marine fish diseases. In: *Fish as Food*, Vol. 2, New York: Academic Press, pp. 541–572.
24 Mattheis, T. (1964). Das Vorkommen von *Vibrio anguillarum* in Ostseefischen. *Zentralblatt für Fischerei N. F. XII*, 259–263.
25 Colorni, A., Paperna, I. and Gordin, H. (1981). Bacterial infections in gilthead sea bream *Sparus aurata* cultured at Elat. *Aquaculture*, **23**, 257–267.
26 Burke, J. and Rodgers, L. (1981). Identification of pathogenic bacteria associated with the occurrence of 'red spot' in sea mullet, *Mugil cephalus* L., in south-eastern Queensland. *J. Fish Dis.*, **3**, 153–159.
27 Grimes, D. J., Stemmler, J., Hada, H., May, E. B., Maneval, D., Hetrick, F. M., Jones, R. T., Stoskopf, M. and Colwell, R. R. (1984). *Vibrio* species associated with mortality of sharks held in captivity. *Microbial Ecol.*, **10**, 271–282.
28 Muroga, K., Takahashi, S. and Yamanoi, H. (1979). Non-cholera *Vibrio* isolated from diseased aya. *Bull. Jap. Soc. Sci. Fish.*, **45**, 829–834.
29 Muroga, K., Jo., Y. and Nishibuchi, M. (1976). Pathogenic *Vibrio* isolated from cultured eels. I, Characteristics and taxonomic status. *Fish Path.*, **12**, 141–145.
30 Muroga, K., Nishibuchi, M. and Jo, Y. (1976). Pathogenic *Vibrio* isolated from cultured eels. II, Physiological characteristics and pathogenicity. *Fish Path.*, **11**, 147–151.
31 Egidius, E., Wiik, R., Andersen, K., Hoff, K. A. and Hjeltner, B. (1986). *Vibrio salmonicida* sp. nov; a new fish pathogen. Int. J. System. Bacteriol., **36**, 518–520.
32 Ransom, D. P., Lannan, C. N., Rohovec, J. S. and Fryer, J. L. (1984). Comparison of histopathology caused by *Vibrio anguillarum* and *Vibrio ordalii* and three species of Pacific salmon. *J. Fish Dis.*, **7**, 107–115.

33 Ransom, D. P. (1978). Bacteriologic, immunologic and pathologic studies of *Vibrio* sp. pathogenic to salmonids. PhD. thesis, Oregon State University, Corvallis.
34 Aoki, T., Egusa, S. and Arai, T. (1974). Detection of R factor in naturally occurring *Vibrio anguillarum* strains. *Antimicrobial Agents and Chemotherapy*, **6**, 534–538.
35 Aoki, T., Kitao T., Itabashi, T., Wada, Y. and Sakai, M. (1981). Proteins and lipopolysaccharides in the membrane of *Vibrio anguillarum*. *Develop. Biol. Standard.*, **49**, 225–232.
36 Chart, H. and Trust, T. J. (1984). Characterization of the surface antigens of the marine fish pathogens, *Vibrio anguillarum* and *Vibrio ordalii*. *Can. J. Microbiol.*, **30**, 703–710.
37 Evelyn, T. P. T. (1984). Immunization against pathogenic vibriosis. In. *Symposium on Fish Vaccination*, ed. P. de Kinkelin, Paris: Off. Int. Epiz. pp. 121–150.
38 Evelyn, T. P. T. and Ketcheson, J. E. (1980). Laboratory and field observations on antivibriosis vaccines. In: *Fish Diseases*, ed. W. Ahne, Berlin: Springer-Verlag, pp. 45–54.
39 Kusuda, R., Kawai, K., Jo. Y., Akizuki, T., Fukunaga, M. and Kotake, N. (1978). Efficacy of oral vaccination for vibriosis in cultured ayu. *Bull. Jap. Soc. Sci. Fish.*, **44**, 21–25.
40 Itami, T. and Kusada, R. (1980). Studies on spray vaccination against vibriosis in cultured ayu. II. Duration of vaccination efficacy and effect of different vaccine preparations. *Bull. Jap. Soc. Sci. Fish.*, **46**, 699–703.
41 Holm, K. O. and Jorgensen, T. (1987). A successful vaccination of Atlantic salmon, *Salmo salar* L. against 'Hitra disease' or coldwater vibriosis. *J. Fish Dis.*, **10**, 85–90.
42 Hayashi, K., Kobayashi, S., Jamata, T., and Ozaki, H. (1964). Studies on the vibrio disease of rainbow trout (*Salmo gairdneri irideus*). II. Prophylactic vaccination against the vibrio-disease. *J. Fac. Fish. Prefect. Univ. Mie*, **6**, 181–191.
43 Fryer, J. L., Nelson, J. S. and Garrison, R. L. (1972). Vibriosis in fish. In: *Progress in Fishery and Food Science*, ed. R. W. Moore, Seattle: University of Washington, Publications in Fisheries, pp. 129–133.
44 Fryer, J. L., Rohovec, J. S., Tebbit, G. L., McMichael, J. S. and Pilcher, K. S. (1976). Vaccination for control of infectious diseases in Pacific salmon. *Fish Path.*, **10**, 155–164.
45 Fryer, J. L., Amend, D. F., Harrell, L. W., Novotony, A. S., Plumb, J. A., Rohovec, J. S. and Tebbit, G. L. (1977). *Development of Bacterins and Vaccines for Control of Infectious Diseases of Fish*. Oregon State University. Sea Grant College Program. Publ. No. ORESU-T-77-012.
46 Fryer, J. L., Rohovec, J. S. and Garrison, R. L. (1978). Immunization of salmonids for control of vibriosis. *Mar. Fish. Review*, **40**, 20–23.
47 Johnson, K. A. and Amend, D. F. (1983). Efficacy of *Vibrio anguillarum* and *Yersinia ruckeri* bacterins applied by oral and anal intubation of salmonids. *J. Fish Dis.* **6**, 473–476.
48 Alexander, J. B., Bowers, D. A. and Shamshoom, S. M. (1981). Hyperosmotic infiltration of bacteria into trout: Route of entry and the fate of the infiltrated bacteria. *Develop. Biol. Standard.*, **49**, 441–445.
49 Smith, P. D. (1982). Analysis of the hyperosmotic and bath methods for fish vaccination – Comparison of uptake of particulate and non-particulate antigens. *Devel. Comp. Immun.*, **Suppl. 2**, pp. 181–186.

50 Egidius, E. and Andersen, K. (1979). Bath immunization – a practical and non-stressing method of vaccinating farmed sea rainbow trout (*Salmo gairdneri* Richardson) against vibriosis. *J. Fish Dis.* **2**, 405–410.
51 Gould, R. W., O'Leary, P. J., Garrison, R. L., Rohovec, J. S. and Fryer, J. L. (1978). Spray vaccination: a method for the immunization of fish. *Fish Path.*, **13**, 63–68.
52 Rosenkvist-Jensen, L. (1982). Results from vaccination of Danish rainbow trout against vibriosis. *Devel. Comp. Immunol.* **Suppl. 2**, pp. 187–191.
53 Harrell, L. W., Etlinger, H. M. and Hodgins, H. O. (1975). Humoral factors important in resistance of salmonid fish to bacterial diseases. I. Serum antibody protection of rainbow trout (*Salmo gairdneri*) against vibriosis. *Aquaculture,* **6**, 211–220.
54 Viele, P., Kerstetter, T. H. and Sullivan, J. (1980). Adoptive transfer of immunity against *Vibrio anguillarum* in rainbow trout, *Salmo gairdneri* Richardson, vaccinated by the immersion method. *J. Fish Biol.*, **17**, 379–386.
55 Antipa, R. and Amend, D. P. (1977). Immunization of Pacific salmon: Comparison of intraperitoneal injection and hyperosmotic infiltration of *Vibrio anguillarum* and *Aeromonas salmonicida* bacterins. *J. Fish. Res. Board Can.*, **34**, 203–208.
56 Croy, T. R. and Amend, D. F. (1977). Immunization of sockeye salmon (*Oncorhynchus nerka*) against vibriosis using the hyperosmotic infiltration technique. *Aquaculture,* **12**, 317–325.
57 Kawai, K., Kusuda, R. and Itami, T. (1981). Mechanisms of protection in ayu orally vaccinated for vibriosis. *Fish Pathol.,* **15**, 257–262.
58 Fletcher, T. C. and White, A. (1973). Antibody production in the plaice after oral and parenteral immunization with *Vibrio anguillarum* antigens. *Aquaculture,* **1**, 417–428.
59 Baudin-Laurencin, F. and Tangtrongpiros, J. (1980). Some results of vaccination against vibriosis in Brittany. In: *Fish Diseases,* ed. W. Ahne, Berlin: Springer-Verlag, pp. 60–68.
60 Amend, D. F. and Johnson, K. A. (1981). Current status and future needs of *Vibrio anguillarum* bacterins. *Develop. Biol. Standard.* **49**, 403–417.
61 Horne, M. T., Tatner, M., McDerment, S. and Agius, C. (1982). Vaccination of rainbow trout, *Salmo gairdneri* Richardson, at low temperatures and the long-term persistence of protection. *J. Fish Dis.,* **5**, 343–345.
62 Håstein, T., and Refsti, T. (1986). Vaccination of Rainbow Trout against vibriosis by injection, dip and bath. *Bull. Eur. Ass. Fish Pathol.* **6**, 45–49.
63 Johnson, K. A., Flynn, J. K. and Amend, D. F. (1982). Onset of immunity in salmonid fry vaccination by direct immersion in *Vibrio anguillarum* and *Yersinia ruckeri* bacterins. *J. Fish Dis.,* **5**, 197–205.
64 Johnson, K. A., Flynn, J. K. and Amend, D. F. (1982). Duration of immunity in salmonids vaccinated by direct immersion with *Yersinia ruckeri* and *Vibrio anguillarum* bacterins. *J. Fish Dis.,* **5**, 207–213.
65 Tatner, M. F. and Horne, M. T. (1983). Susceptibility and immunity to *Vibrio anguillarum* in post-hatching rainbow trout fry, *Salmo gairdneri* Richardson 1836. *Dev. Comp. Immunol.* **7**, 465–472.

7

Vaccination against Enteric Redmouth (ERM)

A. E. Ellis

THE DISEASE

Host and geographic range

ERM is caused by the Gram-negative motile bacterium *Yersinia ruckeri* and has been known in the USA since the 1950s. The first publications on the pathology and isolation of the bacterium from rainbow trout were by Rucker[1] and Ross *et al.*[2] respectively. The bacterium was characterized and named by Ewing *et al.*[3]

ERM is principally a disease of rainbow trout but all salmonid species are now considered potential hosts. Clinical disease has also been reported in coregonid species.[4] By the late 1970s ERM had spread to virtually all rainbow trout producing areas in the USA and Canada and *Y. ruckeri* has also been isolated in Australia.[5,6] Since 1983 ERM has been reported from most western European countries and South Africa[7] in rainbow trout and recently in farmed seawater Atlantic salmon in Norway. It is a serious disease which can result in heavy mortalities.

Pathology

ERM is a subacute to acute systemic infection and takes its name from the characteristic reddening of the mouth and opercula which is caused by subcutaneous haemorrhaging. Other signs include inflammation and erosion of the jaws and palate, haemorrhaging at the base of fins and exophthalmia. Internally, haemorrhages may occur in muscle and intestine which may also contain a yellow fluid.

Epidemiology

The disease is most severe in fingerling rainbow trout at temperatures of 15–18°C. Outbreaks are not common below 10°C. In larger fish the disease is

FISH VACCINATION
ISBN 0-12-237485-1

usually less severe and more chronic. Transmission is through the water but clinical disease generally occurs only after the fish have been exposed to large numbers of the pathogen.[2] However, an asymptomatic carrier infection may be established on exposure to small numbers of the bacteria[8] and clinical disease may then be precipitated by poor husbandry conditions or handling.

The bacterium is regarded as an obligate pathogen though it may survive for several months in sediments.[9] The most likely reservoirs of infection are carrier salmonids. The bacterium can be isolated from kidney and lower intestine of apparently healthy carrier rainbow trout[8] and also from faeces.[10] Certain non-salmonid fish may also harbour the pathogen. Strains (serotype 1) pathogenic for rainbow trout have been isolated from apparently healthy coregonids,[4] burbot, *Lola lola*[11] and goldfish.[12] The isolation from goldfish was amongst a batch of fish imported into Ireland for the aquarist trade, high-lighting the possible spread of fish diseases by ornamental fish species. Animals other than fish may also act as reservoirs of the bacterium such as aquatic invertebrates and even terrestrial mammals, namely the musk rat.[13]

Strains of *Y. ruckeri*

While different isolates of the bacterium are relatively phenotypically homogeneous, there exists serological diversity with five serotypes currently recognized using rabbit antisera. They have been designated Type I (Hagerman) which is the most common and most virulent; Type II (O'Leary) which is relatively avirulent; Type III (Australian) which is avirulent;[14] serovar IV and serovar V.[15,16] The virulence of serovar IV is unknown and that of serovar V is very low.[17]

Serological characteristics of bacteria depend upon the antigenic structures on the surface of the cell. The major surface molecule of Gram-negative bacteria is lipopolysaccharide (LPS). Flett and Stevenson[17] analysed the electrophoretic patterns of the LPS from representatives of serotypes I, II, III and V. There was similarity of the patterns within the serotypes I, III and V and between serotypes I and II but these were distinct from strains of serovar V. The LPS patterns of strains in serotype II were heterogeneous indicating that the serological character of this group must depend upon cross-reactions between other non-LPS antigens.

Limitations of control

Yersinia ruckeri has been controlled by a broad spectrum of antibiotics including sulphamerazine, oxytetracycline, tribrissen and oxolinic acid but antibiotic resistance has developed in many areas.[18] Furthermore, it is commonly found that once antibiotic treatment is withdrawn the disease recurs and further treatments are required. Thus, antibiotic therapy can often be expensive and the effective vaccines which are commercially available are a cheaper and more effective means of control.

ERM VACCINES

Needs of the vaccine

The story of ERM vaccines is, for the most part, one of great success. The disease mainly affects fingerling salmonids in intensive aquaculture systems. Immersion vaccination, which is highly effective, meets the requirements very well for this industry. While the bacterium occurs in several serotypes it is only serotype I (Hagerman) which is highly virulent and vaccines based on this serotype appear to be highly protective to infection by other serotypes.[14]

Nature of the vaccines

The vaccine (or bacterin) is based on whole bacterial cultures grown in tryptic soy broth which are then rendered safe by inactivation, usually with formalin.

The first ERM vaccine tested was an oral vaccine prepared from phenol-killed bacteria incorporated into the feed (15 ml packed *Y. ruckeri* cells in 10 kg diet).[19] Rainbow trout were fed five times/week for 2 weeks, then once a week for the duration of the test. After 70 days immunized fish showed 90 per cent protection to an injected challenge (which killed 90 per cent of control fish) and protection persisted for the 408 days of the test.

Several methods of bacterin production for oral immunization have been tested, including 0.5 and 3 per cent phenol; 3 per cent chloroform; 1 per cent formalin; and sonication of whole bacterial suspensions.[20] All preparations provided protection but the chloroform inactivated bacterins were the most effective.

With development of immersion vaccination against vibriosis in 1977 it soon became evident that this technique was also highly effective for ERM vaccines and much more effective than oral vaccination. Amend *et al.*[21] investigated various parameters of bacterin production to optimize potency for immersion ERM vaccines. Potency of bacterins prepared from tryptic soy broth cultures of *Y. ruckeri* was not affected by pH of the culture in the region of 6.5–7.7 or by culturing the cells for up to 96 h. Inactivation of the cultures using chloroform, formalin, butanol or phenol did not affect potency but for the purposes of vaccine safety and ease of production, formalin was most convenient. The optimum conditions for achieving highest potency of the vaccines were culture of the bacteria at pH 7.2 for 48 h; lysing the cells at pH 9.8 (they lyse spontaneously at this pH) for 1–2 h and inactivation of the lysate with 0.3 per cent formalin. These bacterins are diluted 10 or 20 fold in water and fish are simply immersed in the vaccine for at least 5 s.

Effectiveness in laboratory trials

Many methods of administering ERM vaccines have been tested, namely oral, injection, immersion, spray and anal intubation. In terms of protection the most potent is injection administration and the least is oral.

While oral vaccination, as mentioned above, was effective large amounts of bacterin were used and the vaccine was fed continuously at the rate of once per week.[19] If the rate of oral administration was reduced, the level and duration of protection decreased. Anderson and Nelson[22] fed 1 mg of chloroform-inactivated bacterin/fish for 1 week and protection persisted for less than 6 weeks. On the other hand, fish which were injected with 1 mg of bacterin on a single occasion were protected for over 3 months.

Anal intubation of the bacterin resulted in higher levels of protection than oral delivery.[23] This suggests that the immunizing antigens are destroyed in the foregut and if means could be found of protecting them during passage through this region, oral vaccines may be greatly improved.

Johnson and Amend[24] compared the effectiveness and duration of protection in rainbow trout vaccinated by injection, immersion and spray administration. Their results are shown in Table 7.1. Immersion vaccination provided high levels of protection with considerable duration.

Table 7.1 *Effectiveness of injection, immersion and spray vaccination against ERM rainbow trout weighing 2.2 g (trial 1) or 4 g (trial 2) (Data from Ref. 25.)*

	% mortality, days post-vaccination		
days:	25	76	125
Trial 1			
Injection (i.p.)	2	0	2
Immersion (20 s)	24	13	12
Controls	79	62	59
	% mortality 30 days post-vaccination		
	Vaccinated		*Control*
Trial 2			
Spray (3–5 s)	14		53
Immersion (20 s)	2		66

Factors affecting efficacy of ERM immersion vaccines

These principally involve size of the fish (which is related to the species) and temperature.

Significant levels of protection do not develop in rainbow trout when immersion vaccinated at or below 1 g though the method is effective for chinook salmon of 0.9 g.[25] The duration of immunity is also size dependent. Johnson *et al.*[25] vaccinated rainbow trout of different sizes at 10–14°C. Protection levels fell markedly after 170 days in fish vaccinated when 1.8 g; after 210 days when 3.2 g but protection was retained for at least 300 days when vaccinated at 4.3 g. Thus, the optimum minimum size for vaccinating rainbow trout is about 4 g.

Temperature is known to affect the rate of onset of protection. Rainbow trout were protected within 7 days of immunization at 10°C or 18°C with levels of protection increasing to a maximum by 21 days.[26] However, the minimum temperature at which vaccination is still effective has not been determined. Increasing the exposure time to the vaccine from 5 to 120 s had no effect on efficacy.[26]

Efficacy of field trials

Data from several field trials have been published and all report significant protection. The largest trial and most detailed report is by Tebbit *et al.*[27] who monitored over 22 million vaccinated rainbow trout over a 2 year period in three separate farms. The combined results indicated 84 per cent reduction in mortalities due to ERM. Furthermore, vaccinated fish required 77 per cent less medicated feed and showed 13.7 per cent increase in feed conversion.

Overcoming limitations of present vaccines

It is clear that ERM immersion vaccines are highly effective and commercially useful but there are several areas where further information on the effects of vaccination are still required.

Need for booster vaccination. Protection has been shown to persist for at least 300 days but the rearing time to market may be much longer than this for certain fish e.g. Atlantic salmon. It is possible that low level infection within a farmed population will act as a natural booster maintaining immunity throughout life but in the absence of natural challenge fish may require immunization each year. Seasonal temperature variations may be important in affecting the level of immunity and the minimum temperature allowing effective vaccination is not known.

Effect on carrier status. No information exists as to whether vaccination can eliminate or prevent the carrier state. Such information is important because if vaccinated fish, while being apparently healthy, can still carry the bacterium they could be a serious means of spreading the disease unless methods for detecting carriers and means of restricting their movement are effectively applied.

Serotypic variation. As mentioned earlier, present vaccines are based on the serotype I (Hagerman) strain of *Y. ruckeri* and until now this vaccine appears to be effective against other serotypes which are less virulent. However, with the increased geographical range of salmonid cultivation it is likely that further serotypes will be identified and the present vaccines may be ineffective against them. Vaccines based upon appropriate serotypes would then need to be produced.

Nature of protection. The nature of neither the immunizing antigen nor the immune response is well understood.

Current evidence points to LPS being the major immunizing antigen. LPS can be extracted from bacteria using butanol or phenol. Immersion vaccines prepared from such extracts were virtually as effective as formalin inactivated whole cell bacterins.[21] However, the nature of the antigenic sites on the LPS molecule is unknown. Its identification and monitoring would allow optimization of its production in the preparation of the vaccine and simplify quality control of vaccine batches.

The nature of the protective immune response is still much in doubt. Cossarini-Dunier[28] could find no evidence that serum agglutinating antibody was correlated with protection. Immunity to ERM resulting from vaccination is regarded as specific, principally because of the long duration of protection which implies establishment of immune memory. The mechanism of immunity is, therefore, presumably based upon either secretory antibody, non-agglutinating antibody or cell mediated immunity, or combinations of these. Identification of the immune mechanism would allow a rational approach to further improvements in the preparation and delivery of vaccines in order to optimize the protective response and hence cost effectiveness.

CONCLUSIONS

ERM vaccines have proved their cost effectiveness for large scale commercial use. However, with the recent emergence of the disease in many countries around the world and the recent identification of new serotypes of the bacterium, there is no guarantee that present vaccines will be universally effective. The major problem is the current lack of knowledge concerning the effects of vaccination on the carrier state of ERM. If vaccination does not eliminate the carrier state it may render detection of carrier fish more difficult because overburdened diagnostic laboratories are usually alerted only after clinical signs have been noticed. Transfer of apparently healthy, vaccinated but carrier fish could constitute a major threat of spreading the disease to non-endemic sites.

REFERENCES

1 Rucker, R. R. (1966). Redmouth disease of rainbow trout (*Salmo gairdneri*). *Bull Off. Int. Epizoot.*, **65**, 825–830.

2 Ross, A. J., Rucker, R. R. and Ewing, W. H. (1966). Description of a bacterium associated with redmouth disease of rainbow trout (*Salmo gairdneri*). *Can. J. Microbiol.*, **12**, 763–770.

3 Ewing, W. H., Ross, A. J., Brenner, D. J. and Fanning, G. R. (1978). *Yersinia ruckeri* sp. nov., the redmouth (RM) bacterium. *Int. J. Syst. Bacteriol.*, **28**, 37–44.

4 Rintamaki, P., Valtonen, E. T. and Frerichs, G. N. (1986). Occurrence of *Yersinia ruckeri* infection in farmed whitefish, *Coregonus peled* Gmelin and *Coregonus muksun* Pallas, and Atlantic salmon, *Salmo salar* L., in northern Finland. *J. Fish Dis.*, **9**, 137–140.

5 Bullock, G. L., Stuckey, H. M. and Shotts, E. B. (1978). Enteric redmouth bacterium: comparison of isolates from different geographical areas. *J. Fish Dis.*, **1**, 351–354.

6 Green, M. and Austin, B. (1982). The identification of *Yersinia ruckeri* and its relationship to other representatives of the Enterobacteriaceae. *Aquaculture*, **34**, 185–192.

7 Bragg, R. R. and Henton, M. M. (1986). Isolation of *Yersinia ruckeri* from rainbow trout in South Africa. *Bull. Eur. Ass. Fish Pathol.*, **6**, 5–6.

8 Busch, R. A. and Lingg, A. (1975). Establishment of an asymptomatic carrier state infection of enteric redmouth disease in rainbow trout (*Salmo gairdneri*). *J. Fish. Res. Bd Can.*, **32**, 2429–2432.

9 Austin, B. and Austin, D. A. (1987). *Bacterial Fish Pathogens*, Chichester: Ellis Horwood.

10 Rodgers, C. J. and Hudson, E. B. (1985). A comparison of two methods for isolation of *Yersinia ruckeri* from rainbow trout (*Salmo gairdneri*). *Bull. Eur. Ass. Fish Pathol.*, **5**, 92–93.

11 Dwilow, A. G., Souter, B. W. and Knight, K. (1987). Isolation of *Yersinia ruckeri* from burbot, *Lola lola* (L.), from the MacKenzie River, Canada. *J. Fish Dis.*, **10**, 315–317.

12 McArdle, J. F. and Dooley-Martin, C. (1985). Isolation of *Yersinia ruckeri* type 1 (Hagerman strain) from goldfish *Carassius auratus* (L.). *Bull. Eur. Ass. Fish Pathol.*, **5**, 10–11.

13 Stevenson, R. M. W. and Daly, J. G. (1982). Biochemical and serological characteristics of Ontario isolates of *Yersinia ruckeri*. *Can. J. Fish. Aquat. Sci.*, **39**, 870–876.

14 Bullock, G. L. and Anderson, D. P. (1984). Immunisation against *Yersinia ruckeri*, cause of enteric redmouth disease. In: *Symposium on Fish Vaccination*, ed. De Kinkelin, P. OIE Paris, pp. 151–166.

15 Stevenson, R. M. W. and Airdrie, D. E. (1984). Serological variation among *Yersinia ruckeri* strains. *J. Fish Dis.*, **7**, 247–254.

16 Daly, J. G., Lindvik, B. and Stevenson, R. M. W. (1986). Serological heterogeneity of recent isolates of *Yersinia ruckeri* from Ontario and British Columbia. *Dis. Aquat. Org.*, **1**, 151–153.

17 Flett, D. E. and Stevenson, R. M. W. (1987). Analysis of the specific antigens of *Yersinia ruckeri*, recognised by salmonid fish. *J. Fish Biol.*, **31**, Supplement A.

18 Ceschia, G., Giorgetti, G., Bertoldini, G. and Fontebasso, S. (1987). The *in vitro* sensitivity of *Yersinia ruckeri* to specific antibiotics. *J. Fish Dis.*, **10**, 65–67.

19 Ross, A. J. and Klontz, G. W. (1965). Oral immunisation of rainbow trout (*Salmo gairdneri*) against an etiological agent of 'Redmouth disease'. *J. Fish. Res. Bd Can.*, **22**, 713–719.

20 Anderson, D. P. and Ross, A. J. (1972). Comparative study of Hagerman redmouth disease and bacterins. *Prog. Fish Cult.*, **34**, 226–228.

21 Croy, T. R. and Amend, D. F. (1977). Immunization of sockeye salmon (*Onchorhynchus nerka*) against vibriosis using the hyperosmotic infiltration technique. *Aquaculture*, **12**, 317–325.

22 Amend, D. R., Johnson, K. A., Croy, T. R. and McCarthy, D. H. (1983). Some factors affecting the potency of *Yersinia ruckeri* bacterins. *J. Fish Dis.*, **6**, 337–344.

23 Anderson, D. P. and Nelson, J. R. (1974). Comparison of protection in rainbow trout (*Salmo gairdneri*) inoculated with and fed Hagerman redmouth bacterins. *J. Fish. Res. Bd Can.*, **31**, 214–216.

24 Johnson, K. A. and Amend, D. F. (1983). Efficacy of *Vibrio anguillarum* and *Yersinia ruckeri* bacterins applied by oral and anal intubation of salmonids. *J. Fish Dis.*, **6**, 473–476.

25 Johnson, K. A. and Amend, D. F. (1983). Comparison of efficacy of several delivery methods using *Yersinia ruckeri* bacterin on rainbow trout, *Salmo gairdneri* Richardson. *J. Fish Dis.*, **6**, 331–336.

26 Johnson, K. A., Flynn, J. K. and Amend, D. F. (1982). Duration of immunity in salmonids vaccinated by direct immersion with *Yersinia ruckeri* and *Vibrio anguillarum* bacterins. *J. Fish Dis.*, **5**, 207–213.

27 Johnson, K. A., Flynn, J. K. and Amend, D. F. (1982). Onset of immunity in salmonid fry vaccinated by direct immersion in *Vibrio anguillarum* and *Yersinia ruckeri* bacterins. *J. Fish Dis.*, **5**, 197–205.

28 Tebbit, G. L., Erickson, J. D. and Vande Water, R. B. (1981). Development and use of *Yersinia ruckeri* bacterins to control enteric redmouth disease. *Dev. Biol. Standard.*, **49**, 395–401.

29 Cossarini-Dunier, H. (1986). Protection against enteric redmouth disease in rainbow trout, *Salmo gairdneri* Richardson, after vaccination with *Yersinia ruckeri* bacterin. *J. Fish Dis.*, **9**, 27–33.

8

Furunculosis Vaccines

T. S. Hastings

INTRODUCTION

Furunculosis, caused by the Gram-negative non-motile bacterium *Aeromonas salmonicida*, is one of the most serious infectious diseases of salmonid fish. The disease is named after the raised liquefactive muscle lesions (furuncles) which sometimes occur in chronically infected fish though these lesions are rarely seen in acute infections which are characterized by a rapidly fatal septicaemia.[1] Some devastating epizootics of furunculosis have been recorded in wild fish populations,[2] but the major economic impact of the disease in recent years has been on salmonid cultivation, principally in Europe, North America and Japan.

The host range of *A. salmonicida* also includes many non-salmonid freshwater and marine fish species. Indeed, 'atypical' strains of *A. salmonicida* have recently been linked with serious ulcerative diseases in commercially reared carp,[3,4] goldfish[5,6] and eels[7] as well as salmonid fish.[8] Atypical strains differ biochemically from the typical strains of *A. salmonicida* (subspecies *salmonicida*) which cause furunculosis, but their precise taxonomic position is at present disputed. A separate subspecies (*A. salmonicida* subsp. *masoucida*) has been isolated from salmonid fish in Japan.[9]

The scientific literature on *A. salmonicida* is extensive and only a small part of it can be referred to here, but further information on many aspects can be found in other reviews.[1,8,10–13] Until now efforts to develop vaccines against *A. salmonicida* have largely been directed towards the prevention of furunculosis in salmonids. However, in view of the emerging importance of diseases caused by atypical strains of *A. salmonicida*, prospects for their control by vaccination will be briefly considered.

THE DISEASE PROBLEM

Most if not all species and age groups of salmonid fish may be affected by furunculosis, though some species seem to be more susceptible than others. For example, Atlantic salmon and brown trout are highly susceptible whereas some strains of rainbow trout seem to be remarkably resistant to the disease.[14]

FISH VACCINATION
ISBN 0-12-237485-1

The clinical signs of furunculosis are somewhat variable depending on the form of the disease, but affected fish often show loss of appetite, darkening of the skin and lethargy.[1] Haemorrhaging may occur at the bases of fins and in the abdominal walls, heart and liver. Enlargement of the spleen and inflammation of the lower intestine are common features of chronic infections, but in acute outbreaks fish may die rapidly with few signs.

Furunculosis is primarily a disease of rising and elevated water temperatures, nevertheless mortalities can occur at temperatures as low as 4–6°C. The risk of disease is especially high if fish are overcrowded, but outbreaks can be precipitated by a variety of stressors such as handling, grading, transporting or following transfer of smolts to sea water.

The major route of transmission of furunculosis appears to be lateral i.e. via infected fish and contaminated water. Although *A. salmonicida* has been isolated from the gonads of infected fish, attempts to demonstrate vertical transmission of the bacterium via infected ova have generally been unsuccessful.[15] It is not certain how long *A. salmonicida* can survive in the aquatic environment, and indeed survival is likely to depend on factors such as water temperature, nutrient levels and the presence of other bacteria. Some studies have suggested that the bacterium may remain viable for several weeks in fresh and sea water,[13] and for many months in river sediments.[16] The normal route(s) of entry of *A. salmonicida* into fish is also uncertain. Infection through the gastrointestinal tract or skin is possible, but another route of entry may be through the gills.[17]

Paradoxically, fish sometimes harbour *A. salmonicida* without showing any signs of disease. These asymptomatic carriers, which may be present in wild or farmed populations, can transmit infection to other fish or, when stressed, may themselves succumb to disease. Bacterial numbers in carrier fish can be very low, and their presence is rarely revealed by routine bacteriological methods. However, a sensitive but time-comsuming technique for detecting carriers, devised by Bullock and Stuckey,[18] is based on injection of suspect fish with a corticosteroid and holding them at elevated temperatures in order to induce clinical disease. Using this technique, Scallon and Smith[19] found that 60–100 per cent of Atlantic salmon smolts from one Irish hatchery were asymptomatic carriers of *A. salmonicida*.

The most widely used method of controlling furunculosis in cultivated fish is by use of antibacterial drugs. Oxytetracycline, potentiated sulphonamides and quinolones such as flumequine and oxolinic acid are among the most effective agents. The use of these and other antibiotics in fish has been reviewed by Michel,[20] but some of their limitations should be emphasized.

Antibiotics can be expensive, especially if several treatments are needed during a single season, and lengthy withdrawal times are often necessary prior to sale of fish for human consumption. Most antibacterials are administered in the feed, but furunculosis-affected fish often lose appetite, preventing effective treatment. Many antibiotics inhibit growth of *A. salmonicida* but do not actually kill the bacterium, and fish which have been treated may remain carriers. In any case, fish will be susceptible to reinfection shortly after treatment has been completed. A serious problem in recent years has been the increasing prevalence of drug resistant strains of *A. salmonicida*. Resistance can occur to most of the commonly used antibacterial agents,[21-23] and in some countries strains of *A. salmonicida* with multiple antibiotic resistance are now commonly isolated.

THE IDEAL VACCINE AND ITS STRATEGIC USE

A good furunculosis vaccine should be capable of providing solid, reliable and long-term protection (e.g. up to 2 years for Atlantic salmon production in sea water). It would be advantageous if the vaccine was also effective against atypical strains of *A. salmonicida*.

Asymptomatic carriers of *A. salmonicida* are important reservoirs of infection, so an ideal vaccine would not only prevent clinical disease but also eliminate carrier infections. By eliminating the pathogen, the risk of disease should be reduced, perhaps even after vaccine-induced immunity had declined, provided infection was not reintroduced from an external source.

Ideally the method of administration of the vaccine should be convenient and cost-effective for immunizing large numbers of fish, and in most situations an oral or immersion vaccine would be preferable to one delivered by injection. However, one potential limitation of oral vaccination should be mentioned. In some countries, consideration has been given to establishing vaccination programmes for wild salmon broodstock: these fish do not feed in fresh water and would need to be immunized by other methods, for example by injection or immersion.

It is difficult to generalize on the most suitable times at which to vaccinate fish against furunculosis. For fish reared in pathogen-free water, vaccination should only be necessary if they are to be moved to an infected area. For fish reared in an infected area, critical times to vaccinate might be during periods preceding high water temperatures, grading, transportation or, in salmon, transfer of smolts to sea water.

EARLY VACCINES

The earliest recorded attempt to immunize fish against furunculosis is that of Duff,[24] who administered an oral vaccine consisting of chloroform-killed *A.*

salmonicida cells to cut-throat trout (*S. clarkii*). When exposed to *A. salmonicida* infection, mortalities in vaccinees were consistently lower than those in control unimmunized fish. During the next 35 years many other workers attempted to emulate Duff's success using oral or injectable vaccines based on whole killed cells, but with conflicting results. This largely empirical and disappointing phase of furunculosis vaccine research has been reviewed by Michel[10] and Munro.[11]

VIRULENCE FACTORS OF *A. SALMONICIDA*

It is now widely accepted that antibacterial vaccines work by inducing immunity against virulence factors i.e. those antigens which enable the bacterium to infect its host and cause disease. During the last 10 years, considerable research effort has been devoted to investigating the virulence mechanisms of *A. salmonicida* and its antigenic composition, the results of which have led to several new experimental vaccines. Only a brief outline of progress in this field can be given here, however some aspects have been discussed in reviews by Munro,[10] Trust[12] and Austin and Austin.[13]

It has been known for many years that freshly isolated *A. salmonicida* cells autoagglutinate when suspended in saline. However, if maintained for long periods of time in the laboratory or cultured at elevated temperatures, many strains lose their autoagglutinating properties and also their virulence. The link between these phenomena was first reported by Udey and Fryer,[25] who found that autoagglutination and virulence of *A. salmonicida* were dependent on possession of an 'additional' cell surface layer (A-layer) external to the outer cell membrane. The A-layer was subsequently found to be mainly composed of a 50 kd protein called the A-protein.[26,27] The A-layer may play an important role in enabling *A. salmonicida* to adhere to host cells,[28] acquire iron *in vivo*,[29] and avoid lysis by serum complement[30] and phagocytosis by host macrophages.[31] The A-protein(s) of most, if not all, typical and atypical strains are antigenically closely related.[32]

Lipopolysaccharide (LPS, endotoxin), another major cell envelope antigen, is composed of three moieties: lipid A, a core oligosaccharide and an O-polysaccharide (O-antigen) which is exposed at the cell surface (Fig. 8.1). Like the A-protein, the O-antigen appears to assist *A. salmonicida* to resist the host's normal serum bactericidal mechanisms.[30] Unlike many other Gram-negative bacteria, the O-antigens of both typical and atypical strains of *A. salmonicida* appear to be biochemically homogeneous and antigenically cross-reactive.[33] Evidence for a further polysaccharide (Ps) antigen, distinct from LPS, has also been reported.[32]

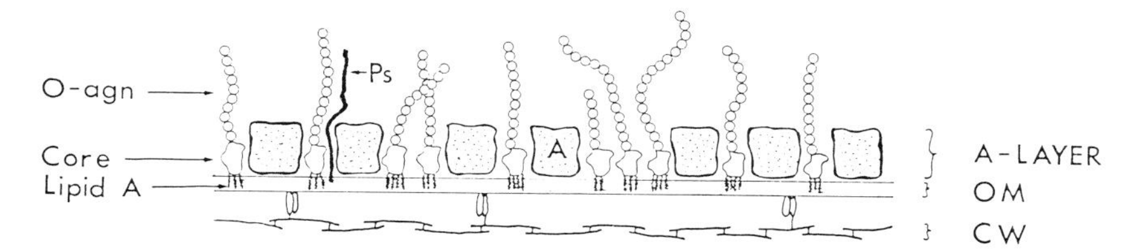

Fig. 8.1 Diagram of the cell surface of *Aeromonas salmonicida*, based on the model of Evenberg *et al.*[32] The outer membrane (OM) and A-layer are external to the peptidoglycan cell wall (CW). The A-layer consists of regularly arranged subunits of the A-protein (A). Lipopolysaccharide (LPS) is composed of an O-antigen, core and a lipid A moiety which is inserted in the outer membrane. An additional polysaccharide (Ps) antigen may also be inserted in the outer membrane.

While cell surface antigens are important in enabling *A. salmonicida* to survive within fish, much of the pathology of furunculosis is attributable to extracellular products (ECP) released during bacterial growth and multiplication.[34,83] The ECPs of typical strains of *A. salmonicida* comprise at least 25 proteins (Fig. 8.2), including a number of enzymes and toxins, as well as other factors (Table 8.1). However, many ECP components have yet to be identified and characterized, including the lethal toxin.[35]

Complex interrelationships have been demonstrated among ECP components,[36] making it difficult to unravel their respective roles in virulence. The extracellular caseinase, a serine protease of molecular weight 70 kd, is believed to play an important role in pathogenicity,[37,38] possibly by digesting host tissues and proteins and providing essential nutrients for bacterial growth.[39] Interestingly, this protease is resistant to most of the protease inhibitors present in normal fish serum, though it is inhibited by α_2-macroglobulin.[40]

Membrane-damaging toxins, which can disrupt a variety of host cells including leucocytes and erythrocytes, are also likely to be important in the pathogenesis of furunculosis, and siderophores may play a vital role in sequestering iron necessary for bacterial growth and multiplication.[41] LPS, normally a cell envelope component, is also released extracellularly.

The nature of ECP of atypical strains of *A. salmonicida* is as yet poorly understood. There is some evidence, however, that they may differ significantly from typical strains with respect to protease, haemolysin and siderophore production.[41,42]

It should be noted that the production of cell-associated and extracellular antigens by *A. salmonicida in vitro* is dependent not only on the isolate used but also on its growth conditions, factors which have important implications for vaccine production.

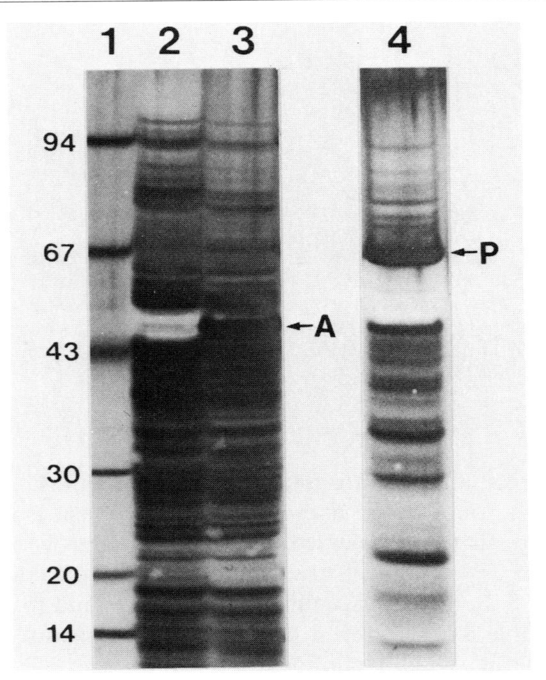

Fig. 8.2 SDS-polyacrylamide gel electrophoresis of whole cell extracts and extracellular products (ECP) of *Aeromonas salmonicida*. Lane 1, molecular weight standards; lane 2, extract of whole cells (A− strain); lane 3, extract of whole cells (A+ strain); lane 4, ECP. Arrows indicate the location of A-protein (A) and serine protease (P) bands. Sizes of protein standards are indicated in kilodaltons at the side of the figure. Proteins were detected by Western blotting and colloidal gold stain.

Table 8.1 *Extracellular products of* Aeromonas salmonicida

Factor	Reference
Proteinases	
Protease (caseinase, serine protease)	73, 34, 38
Collagenase (metallo-protease)	74
Membrane-active toxins	
Leucocytolysin	75
Phospholipase	76
Cytotoxic glycoprotein	54, 37, 48
Enterotoxin?	77
T-lysin (haemolysin)	78, 42
H-lysin (haemolysin)	79, 36
Other factors	
Lipopolysaccharide*	80
Siderophore	33
Brown pigment	81

*Also a cell surface antigen.

NATURE OF RECENT VACCINES

Five basic types of vaccine have been tested during the last 10 years.

Whole killed or disrupted cells (bacterins)

The discovery of the A layer has revived interest in vaccines based on whole killed cells, in particular the efficacy of oral, injectable and immersion vaccines prepared from virulent as opposed to avirulent strains. Some vaccines have been based on whole formalin-killed cells,[43] whereas in others cells have been disrupted by sonication[25] or treatment with EDTA.[44] Vaccines prepared from virulent strains were found to possess an additional antigen lacking in avirulent strains,[43,44] but their antigenic composition has otherwise been undefined.

ECP and ECP-toxoids

As some components of ECP are toxic for fish,[37,38,45] it is potentially hazardous to immunize fish with crude ECP unless it can be inactivated or toxoided. The first formalin-treated ECP-toxoid proved toxic for brown trout when administered orally,[46] but this problem was later overcome using another method of toxoiding.[47] Surprisingly, no adverse effects were reported when brook trout, Atlantic salmon or coho salmon were immunized by injection with active ECP.[43,48,49] However, the antigenic composition of these vaccines was not defined, and some may have been deficient in one or more factors.

Whole killed cells plus ECP

Vaccines comprising whole cells and ECP might be expected to contain a maximum number of different antigens. Two variations of a combined whole cell-plus-ECP vaccine have been tested to date. The first consisted of a whole chloroform-inactivated broth culture of an avirulent strain of *A. salmonicida* administered to fish by immersion.[50] The second, an oral vaccine, contained formalin-killed cells of a virulent strain plus an ECP toxoid.[47]

Live attenuated vaccines

Live vaccines have received relatively little attention, undoubtedly because of fears regarding their safety in the field. One such vaccine was prepared by growing an autoagglutinating virulent strain of *A. salmonicida* in the laboratory for a prolonged period until it became non-autoagglutinating and avirulent.[51]

Purified antigens

Only three antigens of *A. salmonicida* appear to have been tested in purified or partially purified form: protease,[49,52] LPS[53] and a glycoprotein-containing fraction of ECP.[54] While Sakai[52] inactivated partially purified protease with normal trout serum prior to immunizing sockeye salmon, Shieh[49] i.m.-injected Atlantic salmon with active protease. All of these antigens were at least partially purified but the degree of purity of each is unclear and it is possible that some preparations may have contained other bacterial antigens.

Passive immunization

In some cases fish have been passively immunized using antiserum to *A. salmonicida* antigens raised in fish or in mammals. This form of immunization, which can provide short-term protection, has proved useful for demonstrating the protective nature of antibodies raised against different bacterial strains or antigens.[43,44,55]

VACCINE TRIALS

Table 8.2 lists a number of reports on laboratory and field trials of furunculosis vaccines during the last 10 years. Vaccines based on whole killed or disrupted cells of virulent strains of *A. salmonicida* have been most widely evaluated, but to date there seems to be little agreement as to their efficacy. McCarthy *et al.*[44] and Olivier *et al.*[43] found that virulent strains possessed an additional antigen lacking in avirulent strains.

Injection with killed cells or passive immunization with rabbit antibodies to the antigen, believed to be the A-protein, conferred significant protection to coho salmon against experimental challenge. Single immersion vaccination with EDTA-disrupted cells gave variable results, but protection was improved by concurrent hyperosmotic infiltration or administration of boosters.[56–58] Other workers have reported good, variable or no protection in fish immunized by various routes with whole killed or disrupted cells of virulent strains, and it has been claimed that the A-protein, whilst immunogenic, is not a protective antigen.[12] Unfortunately, several large-scale field trials with disrupted cell vaccines failed because the expected natural furunculosis challenges did not materialize and fish suffered infection by other pathogens.[44]

able 8.2 *Vaccine trials*

accine rain[a]	Method of administration*	Adjuvant	Fish species used	Method of challenge	Protection[b]	Reference
'hole/disrupted cell vaccines						
S-70 (V)	oral	$Al(OH)_3$	co	field	−(W)	25
S-70 (V)	i.p.	FCA	co	field	+(W)	25
47R (V)	h.i.		as	field	+(W)	82
	oral		bt	field	+	63
(V)	i.p.		bk	immersion	−	54
S-1R (V)	i.p.		co,ch,ss	immersion	++	44
S-70 (AV)	i.p.		co	immersion	−	44
S-1R (V)	i.p.		co	field	?(W)	44
S-1R (V)	immersion		co	field	?(W)	44
S-1R (V)	immersion		co,ch	immersion	+	57
S-1R (V)	immersion		co	immersion	−	56
S-76-30 (V)	i.p.	FCA	co	i.p.	+	43
S-75-74 (AV)	i.p.	FCA	co	i.p.	(+)	43
S-1 (V)	i.p.		bk,rt	immersion	+	58
S-1 (V)	i.p.		as	immersion	−	58
S-1 (V)	h.i.		as,bk,rt	immersion	+	58
G 51/79 (V)	i.p.		rt	immersion	−	69
CP and ECP-toxoid vaccines						
V)	i.p.		bk	immersion	+	54
95 (AV)	immersion		bk	immersion	−	50
V)	oral		rt	field	−	47
S-76-30 (V)	i.p.	FCA	co	i.p.	(+)	43
hole-cell-plus-ECP vaccines						
95 (AV)	immersion		bk,as	immersion	+	50
95 (AV)	immersion		bt	field	++	50
V)	oral		rt	field	(+)	47
ve attenuated vaccine						
95 (AV)	immersion		bk,as	immersion	+	51
ycoprotein vaccine						
rious	i.p.		bk	immersion	+	48
otease vaccines						
7301 (V)	i.p.		ss	i.p.	+	52
CMB 1102 (AV)	i.m.		as	i.m.	++	49
dotoxin vaccine						
01 (V)	i.p.	FCA	bk	immersion	+	53
mmercial vaccines						
	i.p.		co,as,bk	immersion	++	59
	i.p.		st	field	+	59
	i.p.		as	immersion	−	60
	immersion		as	immersion	(+)	60

(V) = virulent, (AV) = avirulent; (W) = weak challenge.
Protection based on Relative Per Cent Survival (RPS): +++ 100%, ++ >90%, + >50%, (+)
50%. as = Atlantic salmon, bk = brook trout, bt = brown trout, co = coho salmon, ch = chinook
mon, rt = rainbow trout, ss = sockeye salmon, st = speckled trout.
p. = intraperitoneal injection, h.i. = hyperosmotic infiltration, i.m. = intramuscular injection.

Vaccines based on killed cells of avirulent strains of *A. salmonicida* appear to have been generally ineffective. Nevertheless, immersion vaccination of Atlantic salmon and brook trout with live cells of an avirulent strain gave promising results in laboratory trials.[51] To date, field trials with live vaccines have not been attempted.

A relatively small number of trials have so far been attempted with vaccines consisting of crude ECP or ECP-toxoids. The first formalin-treated ECP-toxoid proved toxic for brown trout when administered orally.[46] Although this problem was overcome by using a different method of toxoiding, field trials with the modified vaccine gave disappointing results.[47] Nevertheless, several workers have reported that injection of fish with crude active ECP or purified components of ECP (protease, LPS or a glycoprotein) could confer some protection against furunculosis in laboratory trials, though LPS was reported to confer protection only when administered with Freund's Complete Adjuvant (FCA).[53] There has been marked disagreement, however, as to the merits of using virulent or avirulent strains for vaccine preparation.[43,48,49,54] Oral and immersion vaccines comprising mixtures of whole killed cells and crude ECP or ECP-toxoid have been reported to protect fish in both laboratory and small-scale field trials.[47,50]

Commercially manufactured furunculosis vaccines were first licensed for use in the United States in the early part of this decade and are now marketed in a number of countries. Injectable vaccines are most widely available but immersion vaccines have recently been licensed for use in Canada and are undergoing trials in several other countries. Little information is generally available on the composition of these vaccines and few reports on their efficacy have yet been published in the scientific literature. One vaccine has been reported to confer 87–97 per cent protection to salmonid fish against experimental furunculosis and 73 per cent protection in field trials.[59] However, independent laboratory trials of a commercial vaccine produced by an unnamed manufacturer suggested that the vaccine provided little or no protection to Atlantic salmon.[60]

EFFECTIVENESS

The variability in reported efficacy of some types of vaccine, especially those based on whole killed cells or crude ECP, is striking but may not be entirely surprising. Different workers have used a variety of bacterial strains and methods in vaccine production, factors which could alter the antigenic composition of the vaccines which, in most cases, have been poorly defined. Several other factors might affect the outcome of a trial, including the method of administration, use of adjuvants or boosters, fish species and the nature of the challenge.

Initial trials with combined whole-cell-plus-ECP vaccines, a live attenuated vaccine and vaccines based on purified protease, LPS and glycoprotein have been promising, though it should be remembered that these vaccines have not yet been as widely evaluated as those based on killed cells or crude ECP.

The effectiveness of a vaccine should be measurable not only in terms of protection but also by its ability to elicit a specific immune response in fish to a 'protective' antigen or antigens. To date the most widely used indicator of specific immunity has been the production of agglutinating antibodies against *A. salmonicida*. However, high titres of these antibodies, which appear to be raised predominantly against LPS,[61] do not necessarily correlate with protection against furunculosis,[43,44,62] and most workers have failed to link protection with a specific immune response in vaccinated fish. Interestingly, one group reported evidence of a specific cell-mediated response of brown trout against *A. salmonicida* which appeared to correlate with protection against natural infection.[63]

By virtue of its memory component, specific immunity might be expected to confer reasonable longevity of protection, but in most successful trials fish have been challenged within 3–6 weeks of being vaccinated. It is possible, therefore, that in some cases protection may be attributable to stimulation of non-specific cellular or humoral defences. Indeed, Olivier *et al.*[64,65] found that coho salmon which were injected with modified FCA (only) were considerably more resistant to furunculosis – and vibriosis – than normal fish, apparently due to non-specific activation of macrophages. Subsequent workers have confirmed that macrophages of fish which have been injected with certain immunopotentiators display enhanced phagocytosis and intracellular killing of virulent strains of *A. salmonicida*[66] (C. J. Secombes, pers. comm.). Several workers have claimed that crude or purified ECP components, including protease, can protect fish against furunculosis, but recent reports suggest that many ECP components, including protease, are poorly immunogenic in rainbow trout.[67,68]

LIMITATIONS OF PRESENT VACCINES

Although some vaccines and vaccination procedures have been reported to confer significant protection to fish against furunculosis, none has yet been claimed to completely protect fish in either laboratory or field trials. Several promising vaccines have as yet only been tested in laboratory trials, often only in a single species and age group of salmonid fish, and there has been a general inability to correlate protection with a specific immune response. Where vaccines have been reported to protect fish, the longevity of protection is uncertain. In a majority of laboratory trials, vaccinated fish have been challenged with the same bacterial isolate used in vaccine preparation, and it remains to be shown that protection would be equivalent against other isolates.

In practice it is likely that many vaccinees would be carriers of *A. salmonicida*, but in the absence of experimental data it is not clear whether vaccines would protect such fish to the same extent as non-carriers. Similarly, the ability of vaccines to eliminate carrier infections has rarely been tested, though in one set of trials Michel[69] did find carriers among whole-cell-vaccinated survivors of an experimental furunculosis challenge.

To date, few vaccines have been tested for their ability to protect salmonid or non-salmonid fish against infections caused by atypical strains of *A. salmonicida*. Evenberg *et al.*[70] have conducted preliminary laboratory trials with carp using vaccines prepared from atypical strains of *A. salmonicida*. These workers reported that neither A-protein, LPS nor sublethal infection with an atypical strain was protective, though i.p. injection with crude ECP-toxoid appeared to confer some protection.

OVERCOMING LIMITATIONS

To date, one of the major limiting factors in the development of an effective furunculosis vaccine has been failure to identify antigens of *A. salmonicida* capable of eliciting specific immune protection in fish. However, the identification of such antigens may soon be possible.

Virulence factors are important candidates for the role of protective antigens, but until recently it has proved difficult to establish the precise role and importance of different antigens in virulence. Gene cloning techniques have recently been used to establish the importance of the A-protein in virulence of *A. salmonicida*,[71] and other antigens should be amenable to this approach. Similarly, while Cipriano and Starliper[51] devised a live vaccine based on an A-layer deficient strain of *A. salmonicida*, molecular genetic techniques might be used to construct live attenuated vaccine strains possessing a full range of virulence factors. To date, most studies on the antigenic composition of *A. salmonicida* have been based on bacteria grown *in vitro*. In many cases it remains to be established that putative virulence factors are actually produced during infection. Conversely, some virulence-associated antigens may only be produced *in vivo*.

Passive immunization techniques have proved useful in indicating possible protective antigens. Olivier *et al.*[43] found that rabbit antisera to *A. salmonicida* antigens conferred better protection to coho salmon than antisera raised in fish themselves. This would suggest that some important antigens may be poorly immunogenic in the fish, and indeed several ECP components, including protease, do appear to be poorly immunogenic in rainbow trout.[67] Furthermore, Ellis *et al.*[55] have reported that rabbit antibodies to protease could protect rainbow trout against *A. salmonicida*, indicating that this antigen, if modified to be more immunogenic in the fish, might be an effective vaccine component. Passive immunization with rabbit antisera should prove useful for identifying further potential protective antigens which are otherwise poorly immunogenic in fish.

Many workers have attempted to identify protective antigens using crude or partially purified antigenic preparations, or vaccines of undefined antigenic composition. In order to identify protective antigens it is important that the antigenic composition of vaccines is carefully defined and where possible the most rigorous criteria of purity applied. It is also important to demonstrate specific immune protection in vaccinated fish, and in this respect the role of cell-mediated immunity merits further investigation.

Conflicting reports of vaccine efficacy might be reduced if different workers used the same bacterial strains and methods in vaccine preparation. The major cell envelope antigens, A-protein and LPS, appear to be serologically homogeneous among different strains, but it is not yet known how many extracellular factors are immunologically related among different strains. Certainly, the pathogenesis of ulcerative diseases caused by atypical strains differs from that of typical furunculosis,[72] and this may reflect the production of a different array of virulence factors by those strains.

COMMERCIAL PROSPECTS

Commercial vaccines are now available which have been reported to reduce, but not prevent, furunculosis mortalities in cultivated salmonid fish. Disease outbreaks would still have to be controlled using antibiotics, a factor which must be considered when deciding if vaccination is yet economically worthwhile.

There is every possibility, however, that a vaccine can be developed which will confer solid immunity against furunculosis, but research and development costs are likely to be high. The most appropriate protective antigens need to be identified and their immunogenicity in fish may require to be enhanced, perhaps using biochemical or genetic engineering techniques. Potential vaccines will need to be evaluated for safety, efficacy and longevity of protection in a variety of species and age groups of fish, both carriers and non-carriers, against *A. salmonicida* strains in different countries. More than one vaccine, or a polyvalent vaccine, may prove necessary in order to protect fish against both typical and atypical strains. Economical and reliable methods of scaling-up antigen production will have to be developed, and vaccine production costs will increase if antigen purification or modification proves necessary.

A live attenuated vaccine could face special licensing problems. In particular, its complete safety and inability to revert to virulence would have to be thoroughly proven. Even then, such a vaccine might face severe legal restrictions, for example use only in areas where furunculosis is endemic, construction of vaccines only from indigenous *A. salmonicida* strains, or restrictions on transportation of vaccinated fish.

CONCLUSIONS

Between 1942 and the late 1970s, many empirical attempts to immunize fish using oral or injectable vaccines consisting of killed *A. salmonicida* cells gave disappointing results. During the last 10 years, however, research into the virulence mechanisms and antigenic composition of *A. salmonicida* has led to several new types of experimental furunculosis vaccines.

Passive immunization experiments suggest that the cell envelope A-protein of virulent strains of *A. salmonicida* may be a potential protective antigen, but attempts to vaccinate fish using killed or disrupted cells of virulent strains have yielded conflicting results. Other vaccines – notably combinations of killed cells and ECP, live attenuated vaccines, and vaccines containing partially purified extracellular protease, glycoprotein or LPS – have been reported to confer some protection to salmonid fish. However, it has yet to be shown if any of these vaccines can induce specific long-term immunity against furunculosis, and their effectiveness against atypical strains of *A. salmonicida* is largely unknown.

Nevertheless, several potential protective antigens have been identified, though it may be necessary to enhance their immunogenicity in fish before they can serve as useful components of a vaccine. In the meantime, adjuvants have been shown to enhance the fish's cellular defences against *A. salmonicida*, and it would seem worthwhile to explore further the possible use of immunopotentiators to stimulate short-term protection against furunculosis. This decade has undoubtedly seen the beginnings of a rational approach to vaccine development, but at present the goal of an ideal furunculosis vaccine remains a challenging objective.

REFERENCES

1 McCarthy, D. H. and Roberts, R. J. (1980). Furunculosis of fish – the present state of our knowledge. In: *Advances in Aquatic Microbiology*, ed M. R. Droop and H. W. Jannasch, London: Academic Press, pp. 293–341.

2 Mackie, T. J., Arkwright, J. A., Pryce-Tannatt, Mottram, J. C., Johnston, W. D. and Menzies, W. J. M. (1930, 1933, 1935). *Interim, Second and Final Reports of the Furunculosis Committee*. Edinburgh: HMSO.

3 Bootsma, R., Fijan, N. and Blommaert, J. (1977). Isolation and preliminary identification of the causative agent of carp erythrodermatitis. *Vet. Arch.*, **47**, 291–302.

4 Sovenyi, J. F., Elliott, D. G., Csaba Gy., Olah, J. and Majarnich, J. J. (1984). Cultural, biochemical and serological characteristics of bacterial isolates from carp erythrodermatitis in Hungary. *Rev. sci. tech. Off. int. Epiz.*, **3**, 597–609.

5 Elliott, D. G. and Shotts, E. B. (1980). Aetiology of an ulcerative disease in goldfish *Carassius carassius* (L.): microbiological examination of diseased fish from seven locations. *J. Fish Dis.*, **3**, 133–143.

6 Whittington, R. J., Gudkovs, N., Carrigan, M. J., Ashburner, L. D. and Thurstan, S. J. (1987). Clinical, microbiological and epidemiological findings in recent outbreaks of goldfish ulcer disease due to atypical *Aeromonas salmonicida* in south-eastern Australia. *J. Fish Dis.*, **10**, 353–362.

7 Kitao, T., Yoshida, T., Aoki, T. and Fukudome, M. (1985). Characterization of an atypical *Aeromonas salmonicida* strain causing epizootic ulcer disease in cultured eel. *Fish Pathol.*, **20**, 107–114.

8 Paterson, W. D. (1983). Furunculosis and other associated diseases caused by *Aeromonas salmonicida*. In: *Antigens of Fish Pathogens*, ed. D. P. Anderson, M. Dorson and P. H. Dubourget, Collection Fondation Marcel Merieux, pp. 119–137.

9 Kimura T. (1969). A new subspecies of *Aeromonas salmonicida* as an etiological agent of furunculosis on 'Sakuramasu' (*Oncorhynchus masou*) and pink salmon (*O. gorbuscha*) rearing for maturity. Part 1. On the morphological and physiological properties. *Fish Pathol.*, **3**, 34–44.

10 Michel, C. (1982). Progress towards furunculosis vaccination. In: *Microbial Diseases of Fish*, ed. R. J. Roberts, Society for General Microbiology Special Publication No. 9, London: Academic Press, pp. 151–169.

11 Munro, A. L. S. (1984). A furunculosis vaccine: illusion or achievable objective. In: *Symposium on Fish Vaccination*, ed. P. de Kinkelin, O.I.E., pp. 97–120.

12 Trust, T. J. (1986). Pathogenesis of infectious diseases of fish. *Ann. Rev. Microbiol.*, **40**, 479–502.

13 Austin, B. and Austin, D. A. (1987). *Bacterial Fish Pathogens*. Ellis Horwood Ltd., pp. 111–195.

14 Cipriano, R. C. (1983). Resistance of salmonids to *Aeromonas salmonicida*: relation between agglutinins and neutralizing activities. *Trans. Amer. Fish. Soc.* **112**, 95–99.

15 McCarthy, D. H. (1977). Some ecological aspects of the bacterial fish pathogen – *Aeromonas salmonicida*. In: *Aquatic Microbiology*, ed. F. A. Skinner and J. M. Shewan, Society for Applied Bacteriology Symposium No. 6. London: Academic Press, pp. 294–324.

16 Michel, C. and Dubois-Darnaudpeys, A. (1980). Persistence of the virulence of *Aeromonas salmonicida* strains kept in river sediments. *Ann. Rech. Vet.*, **11**, 375–386.

17 Hodgkinson, J. L., Bucke, D. and Austin, B. (1987). Uptake of the fish pathogen, *Aeromonas salmonicida*, by rainbow trout (*Salmo gairdneri* L.). *FEMS Microbiol. Lett.*, **40**, 207–210.
18 Bullock, G. L. and Stuckey, H. M. (1975). *Aeromonas salmonicida*: detection of asymptomatically infected trout. *Prog. Fish-Cult.*, **37**, 237–239.
19 Scallon, A. and Smith, P. R. (1985). Control of asymptomatic carriage of *Aeromonas salmonicida* in Atlantic salmon smolts with flumequine. In: *Fish and Shellfish Pathology*, ed. A. E. Ellis, London: Academic Press, pp. 119–127.
20 Michel, C. (1986). Practical value, potential dangers and methods of using antibacterial drugs in fish. *Rev. sci. tech. Off. int. Epiz.*, **5**, 659–675.
21 Aoki, T., Kitao, T., Iemura, N., Mitoma, Y. and Nomura, T. (1983). The susceptibility of *Aeromonas salmonicida* strains isolated in cultured and wild salmonids to various chemotherapeutants. *Bull. J. Soc. Sci. Fish.*, **49**, 17–22.
22 Hastings, T. S. and McKay, A. (1987). Resistance of *Aeromonas salmonicida* to oxolinic acid. *Aquaculture*, **61**, 165–171.
23 O'Grady, P., Palmer, R., Rodger, H. and Smith, P. (1987): Isolation of *Aeromonas salmonicida* strains resistant to the quinoline antibiotics. *Bull. Eur. Assoc. Fish Pathol.*, **7**, 43–46.
24 Duff, D. C. B. (1942). The oral immunization of trout against *Bacterium salmonicida*. *J. Immunol.*, **44**, 87–94.
25 Udey, L. R. and Fryer, J. L. (1978). Immunization of fish with bacterins of *Aeromonas salmonicida*. *Marine Fisheries Review*, **40 (3)**, 12–17.
26 Kay, W. W., Buckley, J. T., Ishiguro, E. E., Phipps, B. M., Monette, J. P. L. and Trust, T. J. (1981). Purification and disposition of a surface protein associated with virulence of *Aeromonas salmonicida*. *J. Bacteriol.*, **147**, 1077–1084.
27 Evenberg, D. and Lugtenberg B. (1982). Cell surface of the fish pathogenic bacterium *Aeromonas salmonicida*. II. Purification and characterization of a major cell envelope protein related to autoagglutination, adhesion and virulence. *Biochim. Biophys. Acta*, **684**, 249–254.
28 Trust, T. J., Kay, W. W. and Ishiguro, E. E. (1983). Cell surface hydrophobicity and macrophage association of *Aeromonas salmonicida*. *Current Microbiology*, **9**, 315–318.
29 Kay, W. W., Phipps, B. M., Ishiguro, E. E. and Trust, T. J. (1985). Porphyrin binding by the surface protein array virulence protein of *Aeromonas salmonicida*. *J. Bacteriol.*, **164**, 1332–1336.
30 Munn, C. B., Ishiguro, E. E., Kay, W. W. and Trust, T. J. (1982). Role of surface components in serum resistance of virulent *Aeromonas salmonicida*. *Infect. Immun.*, **36**, 1069–1075.
31 Sakai, D. K. and Kimura, T. (1985). Relationship between agglutinative properties of *Aeromonas salmonicida* strains isolated from fish in Japan and their resistance to mechanisms of host defense. *Fish Pathol.*, **20**, 9–21.
32 Evenberg, D., Versluis, R. and Lugtenberg, B. (1985). Biochemical and immunological characterization of the cell surface of the fish pathogenic bacterium *Aeromonas salmonicida*. *Biochim. Biophys. Acta*, **815**, 233–244.
33 Chart, H., Shaw, D. H., Ishiguro, E. E. and Trust, T. J. (1984). Structural and immunochemical homogeneity of *Aeromonas salmonicida* lipopolysaccharide. *J. Bacteriol.*, **158**, 16–22.
34 Ellis, A. E., Hastings, T. S. and Munro, A. L. S. (1981). The role of *Aeromonas salmonicida* extracellular products in the pathology of furunculosis. *J. Fish Dis.*, **4**, 41–51.

35 Ellis, A. E., Burrows, A. S. and Stapleton, K. J. (1988). Lack of relationship between virulence of *Aeromonas salmonicida* and the putative virulence factors: A-layer, extracellular proteases and extracellular haemolysins. *J. Fish Dis.* (in press).
36 Titball, R. W., Bell, A. and Munn, C. B. (1985). Role of caseinase from *Aeromonas salmonicida* in activation of hemolysin. *Infect. Immun.*, **49**, 756–759.
37 Tajima, K., Takahashi, T., Ezura, Y. and Kimura, T. (1983). Studies on the virulent factors produced by *Aeromonas salmonicida*, a causative agent of furunculosis in salmonidae II. Studies on the pathogenicity of the protease of *Aeromonas salmonicida* Ar-4 (EFDL) on yamabe (*Oncorhynchus masou* f. *ishikawai*) and goldfish (*Carassius auratus*), and the substance which exhibits cytotoxic effect in RTG-2 (rainbow trout gonad) cells. *Bull. Fac. Fish. Hokk. Univ.*, **34**, 111–123.
38 Sakai, D. K. (1985). Loss of virulence in a protease-deficient mutant of *Aeromonas salmonicida*. *Infect. Immun.*, **48**, 146–152.
39 Sakai, D. K. (1985). Significance of extracellular protease for growth of a heterotrophic bacterium, *Aeromonas salmonicida*. *Appl. Environ. Microbiol.*, **50**, 1031–1037.
40 Ellis, A. E. (1987). Inhibition of the *Aeromonas salmonicida* extracellular protease by α_2-macroglobulin in the serum of rainbow trout. *Microbial Pathogenesis*, **3**, 167–177.
41 Chart, H. and Trust, T. J. (1983). Acquisition of iron by *Aeromonas salmonicida*. *J. Bacteriol.*, **156**, 758–764.
42 Hastings, T. S. and Ellis, A. E. (1985). Differences in production of haemolytic and proteolytic activities by various isolates of *Aeromonas salmonicida*. In: *Fish and Shellfish Pathology*, ed. A. E. Ellis, London: Academic Press, pp. 69–77.
43 Olivier, G., Evelyn, T. P. T. and Lallier, R. (1985a). Immunogenicity of vaccines from a virulent and an avirulent strain of *Aeromonas salmonicida*. *J. Fish Dis.*, **8**, 43–55.
44 McCarthy, D. H., Amend, D. F., Johnson, K. A. and Bloom, J. V. (1983). *Aeromonas salmonicida*: determination of an antigen associated with protective immunity and evaluation of an experimental bacterin. *J. Fish Dis.*, **6**, 155–174.
45 Munro, A. L. S., Hastings, T. S., Ellis, A. E. and Liversidge, J. (1980). Studies on an ichthyotoxic material produced extracellularly by the furunculosis bacterium *Aeromonas salmonicida*. In: *Fish Diseases: Third COPRAQ-Session*, ed. W. Ahne, Berlin: Springer-Verlag, pp. 98–106.
46 Austin, B. and Rodgers, C. J. (1981). Preliminary observations on *Aeromonas salmonicida* vaccines. *Develop. biol. Standard.*, **49**, 387–393.
47 Rodgers, C. J. and Austin, B. (1985). Oral immunization against furunculosis: and evaluation of two field trials. In: *Fish Immunology*, ed. M. J. Manning and M. F. Tatner, London: Academic Press, pp. 185–194.
48 Cipriano, R. C. (1982). Immunization of brook trout (*Salvelinus fontinalis*) against *Aeromonas salmonicida*: immunogenicity of virulent and avirulent isolates and protective ability of different antigens. *Can. J. Fish. Aquat. Sci.*, **39**, 218–221.
49 Shieh, H. S. (1985). Vaccination of Atlantic salmon, *Salmo salar* L., against furunculosis with protease of an avirulent strain of *Aeromonas salmonicida*. *J. Fish Biol.*, **27**, 97–101.
50 Cipriano, R. C., Morrison, J. K. and Starliper, C. E. (1983). Immunization of salmonids against the fish pathogen, *Aeromonas salmonicida*. *J. World Maricul. Soc.*, **14**, 201–211.

51 Cipriano, R. C. and Starliper, C. E. (1982). Immersion and injection vaccination of salmonids against furunculosis with an avirulent strain of *Aeromonas salmonicida*. *Prog. Fish-Cult.*, **44**, 167–169.
52 Sakai, D. K. (1985). Efficacy of specific antibody against agglutinating *Aeromonas salmonicida* strains on infectivity and vaccination with inactivated protease. *J. Fish Dis.*, **8**, 397–405.
53 Cipriano, R. C. and Pyle, S. W. (1985). Adjuvant-dependent immunity and the agglutinin response of fishes against *Aeromonas salmonicida*, cause of furunculosis. *Can. J. Fish. Aquat. Sci.*, **42**, 1290–1295.
54 Cipriano, R. C. (1982). Immunogenic potential of growth products extracted from cultures of *Aeromonas salmonicida* for brook trout (*Salvelinus fontinalis*). *Can. J. Fish. Aquat. Sci.*, **39**, 1512–1518.
55 Ellis, A. E., Burrows, A. S., Hastings, T. S. and Stapleton, K. J. (1988). Identification of *Aeromonas salmonicida* extracellular protease as a protective antigen against furunculosis by passive immunization. *Aquaculture*, **70**, 207–218.
56 Amend, D. F. and Johnson, K. A. (1984). Evidence for lack of antigenic competition among various combinations of *Vibrio anguillarum, Yersinia ruckeri, Aeromonas salmonicida* and *Renibacterium salmoninarum* bacterins when administered to salmonid fishes. *J. Fish Dis.*, **7**, 293–300.
57 Johnson, K. A. and Amend, D. F. (1984). Potential for immersion vaccination against *Aeromonas salmonicida*. *J. Fish Dis.*, **7**, 101–105.
58 Newman, S. G. and Majnarich, J. J. (1985). Immunization of salmonids against furunculosis. *Fish Pathol.*, **20**, 403–411.
59 Paterson, W. D., Lall, S. P., Airdrie, D., Greer, P., Greenham, G. and Poy, M. (1985). Prevention of disease in salmonids by vaccination and dietary modification. *Fish Pathol.*, **20**, 427–434.
60 Michel, C. (1986). Interet et limites des vaccinations antibacteriennes en pisciculture. *La Pisciculture Francaise*, **85**, 4–9.
61 Paterson, W. D. and Fryer, J. L. (1974). Effect of temperature and antigen dose on the antibody response of juvenile Coho salmon (*Oncorhynchus kisutch*) to *Aeromonas salmonicida* endotoxin. *J. Fish. Res. Board Can.*, **31**, 1743–1749.
62 Michel, C. and Faivre, B. (1982). Occurrence and significance of agglutinating antibodies in experimental furunculosis of rainbow trout, *Salmo gairdneri* Richardson. *J. Fish Dis.*, **5**, 429–432.
63 Smith, P. D., McCarthy, D. H. and Paterson, W. D. (1980). Further studies on furunculosis vaccination. In: *Fish Diseases: Third COPRAQ-Session*, ed. W. Ahne, Berlin: Springer-Verlag, pp. 113–118.
64 Olivier, G., Evelyn, T. P. T. and Lallier, R. (1985). Immunity to *Aeromonas salmonicida* in coho salmon (*Oncorhynchus kisutch*) induced by modified Freund's complete adjuvant: its non-specific nature and the probable role of macrophages in the phenomenon. *Dev. Comp. Immunol.*, **9**, 419–432.
65 Olivier, G., Eaton, C. A. and Campbell, N. (1986). Interaction between *Aeromonas salmonicida* and peritoneal macrophages of brook trout (*Salvelinus fontinalis*). *Vet. Immunol. Immunopathol.*, **12**, 223–234.
66 Kitao, T. and Yoshida, Y. (1986). Effect of an immunopotentiator on *Aeromonas salmonicida* infection in rainbow trout (*Salmo gairdneri*). *Vet. Immunol. Immunopathol.*, **12**, 287–296.
67 Hastings, T. S. and Ellis, A. E. (1988). The humoral immune response of rainbow trout and rabbits to extracellular products of *Aeromonas salmonicida*. *J. Fish Dis.*, **11**, 147–160.

68 Ellis, A. E., Stapleton, K. J. and Hastings, T. S. (1988). The humoral immune response of rainbow trout (*Salmo gairdneri*) immunized by various regimes and preparations of *Aeromonas salmonicida* antigens. *Vet. Immunol. Immunopathol.*, (in press).
69 Michel, C. (1985). Failure of anti-furunculosis vaccination of rainbow trout (*Salmo gairdneri* Richardson), using extra-cellular products of *Aeromonas salmonicida* as an immunogen. *Fish Pathol.*, **20**, 445–451.
70 Evenberg, D., de Graff, P., Fleuren, W. and van Muiswinkel, W. B. (1986). Blood changes in carp (*Cyprinus carpio*) induced by ulcerative *Aeromonas salmonicida* infections. *Vet. Immunol. Immunopathol.*, **12**, 321–330.
71 Belland, R. J. and Trust, T. J. (1987). Cloning of the gene for the surface array protein of *Aeromonas salmonicida* and evidence linking loss of expression with genetic deletion. *J. Bacteriol.*, **169**, 4086–4091.
72 Bullock, A. M. and Roberts, R. J. (1980). Inhibition of epidermal migration in the skin of rainbow trout *Salmo gairdneri* Richardson in the presence of achromogenic *Aeromonas salmonicida*. *J. Fish Dis.*, **3**, 517–524.
73 Fyfe, L., Coleman, G. and Munro, A. L. S. (1987). Identification of major common extracellular proteins secreted by *Aeromonas salmonicida* strains isolated from diseased fish. *Appl. Environ. Microbiol.*, **53**, 722–726.
74 Sheeran, B., Drinan, E. and Smith, P. R. (1984). Preliminary studies on the role of extracellular proteolytic enzymes in the pathogenesis of furunculosis. In: *Fish Diseases: Fourth COPRAQ-Session*, ed. ACUIGRUP, Madrid: Editora ATP, pp. 89–100.
75 Fuller, D. W., Pilcher, K. S. and Fryer, J. L. (1977). A leucocytolytic factor isolated from cultures of *Aeromonas salmonicida*. *J. Fish. Res. Board Can.*, **34**, 1118–1125.
76 Buckley, J. T., Halasa, L. N. and MacIntyre, S. (1982). Purification and partial characterization of a bacterial phospholipid: cholesterol acyltransferase. *J. Biol. Chem.*, **257**, 3320–3325.
77 Jiwa, S. F. H. (1983). Enterotoxigenicity, haemagglutination and cell surface hydrophobicity in *Aeromonas hydrophila, A. sobria* and *A. salmonicida*. *Vet. Microbiol.*, **8**, 17–34.
78 Titball, R. W. and Munn, C. B. (1983). Partial purification and properties of a haemolytic activity (T-lysin) from *Aeromonas salmonicida*. *FEMS Microbiol. Lett*, **20**, 207–210.
79 Titball, R. W. and Munn, C. B. (1985). The purification and some properties of H-lysin from *Aeromonas salmonicida*. *J. Gen. Microbiol.*, **131**, 1603–1609.
80 MacIntyre, S., Trust, T. J. and Buckley, J. T. (1980). Identification and characterization of outer membrane fragments released by *Aeromonas* sp. *Can. J. Biochem.*, **58**, 1018–1025.
81 Donlon, J., McGettigan, S., O'Brien, P. and O'Carra, P. (1983). Re-appraisal of the nature of the pigment produced by *Aeromonas salmonicida*. *FEMS Microbiol. Lett.* **19**, 285–290.
82 Palmer, R. and Smith, P. R. (1980). Studies on vaccination of Atlantic salmon against furunculosis. In: *Fish Diseases: Third COPRAQ-Session*, ed. W. Ahne, Berlin: Springer-Verlag, pp. 108–112.
83 Klontz, G. W., Yasutake, W. T. and Ross, A. J. (1966). Bacterial diseases of the salmonidae in the western United States: pathogenesis of furunculosis in rainbow trout. *Am. J. Vet. Res.*, **27**, 1455–1460.

9

Vaccination against *Aeromonas hydrophila*

Roselynn M. W. Stevenson

THE DISEASE PROBLEM

Significance and populations at risk

Aeromonas hydrophila is a ubiquitous and heterogeneous organism which produces disease under stress conditions or in concert with infection by other pathogens. Acute bacterial haemorrhagic septicaemia (motile aeromonad septicaemia) is a systemic disease, which may produce swelling of the body cavity and haemorrhage of organs. In other cases, there may be mortalities without external indications. Damage may also occur as localized infections at sites of injury or parasite attachment. Infection with *A. hydrophila* may be the final insult to fish already affected by parasites, injuries, or other pathogens. For example, red sore disease losses of largemouth bass were attributed to a combination of *A. hydrophila* and the ciliate *Epistylis*,[1] while furunculosis lesions on goldfish can be secondarily infected with *A. hydrophila*.[2] Particularly in warmwater aquaculture, *A. hydrophila* is considered a major economic problem, but it is difficult to distinguish direct losses and those from secondary infections.[3,4]

A. hydrophila is frequently associated with disease in carp, eels, milkfish, channel catfish, tilapia, and ayu.[3,5,6] In salmonids, *A. hydrophila* produces stress-related diseases, particularly when water temperatures rise. Groberg *et al.*[7] demonstrated that the outcome of experimental infections of juvenile steelhead trout, coho salmon and spring chinook salmon was related to water temperature. Mortalities were generally highest in fish held at 20.5°C and 17°C, moderate at 15°C and 12.2°C, and very low or absent at 9.4°C and below. Increased susceptibility may be due to physiological differences in fish at different temperatures. Rapid temperature changes increase stress and the severity of the stress response. For example, a recent epizootic in fingerling rainbow trout in Spain occurred when water temperatures increased from between 5.5°C and 8°C to 11°C.[8] Similarly, motile aeromonad infections of largemouth bass in reservoirs occur with increased water temperatures.[1] Other environmental factors such as crowding, or organic material in the water may contribute to infections by *A. hydrophila*.

FISH VACCINATION
ISBN 0-12-237485-1

Transmission and carriers

A. hydrophila is constantly present in water environments and as part of the normal microbiota of fish. More aeromonads are found when water temperatures are warmer[9] but it is not clear whether outbreaks of disease are related to the levels of the organism in the water or to the physiological effects of fish of higher temperatures.[1] If the total load of aeromonads in natural waters is considered a potential source of disease, the assumption is that environmental isolates and disease isolates are the same, which may not be valid (see below). Many isolates are enterotoxigenic, and thus potential enteric pathogens of humans.[9,10] The presence of high levels of aeromonads in intensive fish culture, such as the sewage-fed fish ponds in Hungary and India,[11] may present a risk of infection not only for fish, but also for human handlers and consumers.

Pathogen heterogeneity

In the past, several synonyms have been used for *A. hydrophila* including *A. liquefaciens, A. punctata,* and *Bacterium punctata.* Bacteria included in the species *A. hydrophila* are heterogeneous biochemically, genetically, serologically, and phenotypically. Differences in relative virulence of strains of *A. hydrophila* is the form of heterogeneity with most significance for fish vaccine development. De Figuerredo and Plumb[12] found environmental and disease isolates differed in virulence when injected into channel catfish. LD_{50} values for isolates from diseased fish were 6.4×10^4 cfu, compared with 1.5×10^6 cfu for environmental isolates. In general, virulent isolates of *A. hydrophila* are considered to be those with LD_{50} values of 10^4 to 10^5, while strains which do not kill fish at 10^7 are considered non-virulent. Values for salmonids,[7,13] carp[14] and other fish appear generally similar to those for channel catfish,[12] although LD_{50} values are varyingly expressed as doses for the animals used or for a specified weight. In almost all cases reported, an injected challenge was used. Both Groberg *et al.*[7] and Ruangapan *et al.*[4] comment on the need for an injection challenge in order to achieve reliable and reproducible infections. On the other hand, Thune, 1980, cited by Newman[5] was able to obtain 45 per cent mortalities when channel catfish fry were exposed to a water borne challenge in laboratory vaccine trials. In vaccine efficacy tests, the challenge to be used may, in itself, pose a major difficulty.

Surface antigens and virulence

Characteristics of bacterial strains may permit prediction of their relative virulence. For *A. hydrophila*, the phenotypic features with the strongest link to virulence are such cell-surface characteristics as cell aggregation and serum-resistance.[13] (The characteristics noted by Mittal *et al.*[13] are frequently cited in the literature as indicative that a strain being studied is virulent.) Leblanc *et al.*[15] serogrouped 195 isolates of *A. hydrophila* based on heat-stable and heat-labile antigens, and concluded that all the virulent strains belonged to the 'LL1 serogroup'. However, there is still no well-defined antigenic scheme for serotyping strains of *A. hydrophila.*[16] In virulent, autoaggregating strains of *A. hydrophila*, the polysaccharide chains of the lipopolysaccharide, O-antigen, are of homogeneous length, and are highly immunogenic.[17] In addition, a surface protein array, the S-layer, occurs.[18] Thus, in two key surface features, virulent isolates of *A. hydrophila* resemble virulent *A. salmonicida*. These characteristics are potentially very significant in resolving questions related to serological heterogeneity and variable virulence of strains of *A. hydrophila*, and in identifying potential vaccine antigens.

Toxins and extracellular products

Motile aeromonads which showed high virulence for rainbow trout,[19] carp, and loach[14] all belonged to *A. hydrophila* biovar hydrophila, rather than the related species *A. caviae* or *A. sobria*. These species are distinguished by a number of biochemical reactions.[16,20] Isolates of *A. hydrophila* from human and fish sources produce a perplexing variety of toxins, including enterotoxin and two cytolytic toxins (α- and β-haemolysins).[21,22] Although Boulanger *et al.*[19] demonstrated enterotoxigenic activities in fish isolates of *A. hydrophila*, it is not clear that the same toxins are significant for both fish and human isolates.

Motile aeromonads produce a variety of extracellular products (ECPs), which include toxins, heterogeneous arrays of proteases,[23] and other enzymes. Gelatinase and caseinase activity appeared to correlate with virulence in studies by Kuo, cited by Wakayabashi *et al.*[14] Hsu *et al.*[24] demonstrated that virulent strains produced large amounts of elastase, caseinase, and staphylolysin. Chondroitinase activity has also been noted as an indicator of virulent strains.[20] When injected into fish, ECPs can produce mortalities and gross pathological tissue changes that suggest they may play a role in the disease process.[14,25,26] Attempts have been made to identify the specific components responsible for toxicity. Allan and Stevenson[25] demonstrated that the products of a protease-deficient mutant were still toxic when injected into rainbow trout, and mortalities were proportional to haemolytic activity injected. Lallier *et al.*[27] attributed toxicity to an unidentified product unique to virulent strains. In channel catfish studies, the effects have been linked variously to two proteases, and β-haemolysin.[22,28] Studies involving injection of ECPs are, however, measures of toxicity, not *in vivo* pathogenesis. Bacterial strains carrying mutations in specific virulence factors provide a means of assessing the contribution of that factor to the invasion and survival of *A. hydrophila* in fish. Leung[29] has used this approach to demonstrate that a transposon-induced protease-deficient mutant of a virulent parent strain of *A. hydrophila* was less virulent and less able to persist when injected intramuscularly into rainbow trout.

If ECPs play a significant part in invasion by the pathogen, then they may also be important antigens in vaccines, capable of inducing effective protection. Some ECPs are antigenically similar in many strains, as, for example, the proteases examined by Leung.[29] Such common antigens may be the means of inducing cross-protection against strains with heterogeneous surface antigens.

Methods of control

The best prevention and treatment for motile aeromonad septicaemia is good husbandry.[30] But, when resources are being used near their limit, it may not be possible to improve conditions. In these situations, antibiotics, particularly oxytetracycline or the nitrofurans, are used for both treatment and prophylaxis. However, extensive use of antibiotics has the serious drawback of increasing plasmid-encoded antibiotic resistances in *A. hydrophila.* Aoki *et al.*[31] found that, of 250 isolates from Japanese fish culture operations, 78 were resistant to one or more common antibiotics, including tetracycline, sulphanilamide, streptomycin, chloramphenicol and the nitrofurans. As *A. hydrophila* is an opportunistic pathogen of humans as well as fish, antibiotic-resistant strains may become a human health problem, particularly when fish is eaten raw[31] or improperly cooked.[32] The problems are not confined to food fish, as baitfish and pet fish[33] can also be a source of antibiotic-resistant *A. hydrophila.* Concerns about extensive use of antibiotics in agriculture and aquaculture make vaccines against *A. hydrophila* an attractive option when there is limited flexibility to improve environmental and management conditions. Although Wu *et al.*[34] proposed use of bacteriophage lysis as a control measure, problems with resistance and host range of phages make this impractical.

THE IDEAL VACCINE AND ITS STRATEGIC USE

A vaccine against motile aeromonad diseases would only be cost-effective in fish culture where there was a recurrent disease problem. The antigenic diversity of *A. hydrophila* strains presents a major problem in vaccine development, and Plumb[35] warned that a polyvalent vaccine might have to contain antigens representing all of the strains that fish might encounter. If, however, virulent strains share significant common antigens, these may be adequate to induce protection against most strains. The surface array observed by Dooley *et al.*[18] would be one candidate antigen. For salmonid culture, a vaccine for the major problem of furunculosis (*A. salmonicida*) which contained sufficient cross-reactive antigens to protect against the more spasmodic outbreaks of bacterial haemorrhagic septicaemia would be of value. In this regard, a similarity in LPS molecules has been noted.[17]

A quite different approach to protection against *A. hydrophila* may be to increase non-specific, natural immune mechanisms. This would be attractive in cases in which aeromonad septicaemia is a problem in concurrent infections with other pathogens. Several studies demonstrate the action of such immunity to *A. hydrophila*. Lamers and Van Muiswinkel[36] found that carp had natural agglutinin titres against heat-killed cells of *A. hydrophila*. The active component did not react with monoclonal antibodies to carp immunoglobulin, indicating that the natural agglutinins were non-immunoglobulin molecules, not antibodies induced by prior exposure to *A. hydrophila*. Extract from the tunicate, *Ecteinascidia turbinata*, acts as an immunopotentiator and increases survival of eels injected with *A. hydrophila*, apparently by enhancing phagocytosis.[37]

NATURE OF PRESENT VACCINES

At present, no vaccines for protection against *A. hydrophila* are commercially available. Bacterins have been prepared by inactivating strains isolated from the specific facility to be treated. The choice of the bacterial strain is important (see pathogen heterogeneity above) and there is some contention that vaccinated fish are not protected against challenge with heterologous isolates of *A. hydrophila*,[35] though observations from protection trials are equivocal.

For bacterin preparation, antigens responsible for inducing the protective response need to be maintained during inactivation steps. Several studies demonstrate that the method of preparation of *A. hydrophila* bacterin significantly influences the immune response. For example, in carp, Lamers and Van Muiswinkel[36] found that vaccines prepared from heated and disrupted cells induced a higher agglutinating antibody titre over the 8-month test period than did formalin-killed cells. The antisera produced in response to formalin-treated and heated antigens cross-reacted, but were not identical. Antibodies produced in response to the formalin-killed antigen reacted with sheep red blood cells coated with bacterial lipopolysaccharide, indicating this was a major antigen, while antibodies induced by heated cells did not. Lamers and Van Muiswinkel[36] suggested that formalin treatment altered *A. hydrophila* antigen structure and, consequently, processing of the antigens by macrophages, while heating and breakage could release more antigenic material. The general conclusion is that attention must be paid to the preparation of the antigen for vaccinations, both with regard to the heterogeneity of the strains and the actual preparation of the bacterin.

VACCINE TRIALS

In many vaccination studies, the question addressed was whether or not fish responded immunologically, using such indicators as antibody titres. Vaccination has rarely been assessed in terms of protection, perhaps because of difficulties in providing an effective challenge, particularly for field trials. For example, in the single literature report of a field trial, the natural challenge was lost in the last year of testing when nutritional and environmental conditions were improved.[30] This work, on rainbow trout, generally suggests that both vaccination and improvements in management and environment can be useful in reducing losses from motile aeromonad septicaemia and improving growth of fish. The report gives little specific data, and the results are complicated by poor challenges in some cases and by such management intervention as medication of feeds.

Reports of other trials, conducted on a laboratory scale, are also complex to evaluate and compare because of significant differences in such things as the vaccine preparation used, the route of exposure, the method of challenge (or whatever other measure of protection was used), and the age, species and group size of fish used. In some cases, insufficient information has been given. Furthermore, most trials have not assessed the specificity of protection or its duration beyond a few weeks. Thus, where trials have indicated protection, it is not clear whether it is due to specific immunity, and therefore potentially long-lived, or due to non-specific mechanisms with no long-lived memory component.

Schäperclaus (1954, 1970, cited by Lamers[38]) first demonstrated protection of carp against *A. hydrophila*, measured as increased survival of laboratory fish. Post[39] provided comprehensive information about protection in rainbow trout, evaluated routes of exposure to heat-killed cells, and attempted to correlate protection with antibody titres and phagocytic indexes. Injection of a strain of *A. hydrophila* that was avirulent to the test strain of rainbow trout produced protection, which is of interest in view of concerns about antigen heterogeneity and specificity. Post[39] and Khalifa and Post[40] evaluated the effect of bacterins applied with adjuvant. This enhanced the antibody response and Post[39] suggested that rainbow trout with antibody titres of 1:64 or higher could be considered immune. Antibody titres induced in channel catfish were also higher when adjuvant was included in the antigen injection and, while antibody titres do not necessarily indicate protective immunity, Plumb[35] reported that in further trials, these fish were protected against a water-borne challenge by a homologous strain of *A. hydrophila*.

Ruangapan *et al.*[4] vaccinated Nile tilapia (*Tilapia nilotica*) with formalin-killed *A. hydrophila*, by i.p. injection. At 1 week post-vaccination, some degree of protection against an injected challenge was observed; between 2 and 5 weeks, there were no mortalities in vaccinated groups, compared with 73–80 per cent in the controls. This trial suggests that vaccination protection may be feasible for tilapia, though the short interval between vaccination and challenge provides no information on the role of specific immunity and duration of protection.

EFFECTIVENESS – THE IMMUNE RESPONSE TO *AEROMONAS HYDROPHILA*

Lamers[38] used *A. hydrophila* as a model antigen in carp, to examine the effects of routes of exposure and antigen preparation on antigen processing in lymphoid tissue and on production of antibody. Although these tests did not examine protection, the results provide background knowledge of the immune response of fish to *A. hydrophila* which is essential for vaccine development.

Antigen doses and routes of administration are of obvious concern in developing a vaccine. In carp, a high dose of injected antigen was required in order for agglutinating antibody titres to increase soon after vaccination, or for long-term maintenance of increased titres. With 10^5 cells injected per 100 g body weight, the primary antibody titres were low, and the memory response to booster doses at intervals of 1 to 12 months were generally not significantly different from the primary response.[41] Vaccination at low antigen levels might induce only weak responses, and poor memory. The best memory response occurred when both the priming and the challenge doses were high. In general, optimal effects were obtained with matching priming and second doses – both high or both low.[41] If this same pattern is followed in the protective response as for antibody production, consideration will need to be given to the level of natural challenge expected in the field. In most tests, vaccines for *A. hydrophila* have been administered by injection followed by an injection challenge. Carp exposed to *A. hydrophila* bacterin by immersion did show an antibody titre increase, but only when given a second bath immersion 1 to 3 months later,[42] and only when the second exposure to the antigen was also by immersion. Injected booster doses produced no greater response (i.e. no memory) compared with naive fish injected with the same dose.[42] Bath immersion with *A. hydrophila* could induce some immune memory, but only after a prolonged period (1–3 months), and only for a relatively short period. Lamers *et al.*[42] suggest that this low response is not surprising as the 0.01 to 0.2 per cent of vaccine concentration taken up by the fish would correspond to an injected dose of 10^4 to 10^5 cells. When injected, a dose of this size induced a weak or negligible response in previous studies.

Antigen localization was two-phase. The first phase was non-specific and the second stage, localization to the melanomacrophage centres of tissues, occurred with the onset of antibody production.[43] Melanomacrophage centres appear to be the site of long-term antigen localization. *A. hydrophila* antigens remained in lymphoid tissue for 12 months after fish were injected with 10^9 cells.[43] Of potential interest in the production of bacterins is the apparent difference in localization of soluble and particulate antigens in fish. The work of Lamers[38] has provided good basic information about the immune response to *A. hydrophila*. In common with all other vaccine preparations, the need is for more information about the protective aspects of the response, which may also involve cell-mediated immune mechanisms.

LIMITATIONS

The heterogeneity of the strains which are potentially disease agents is generally given as the major problem limiting development of vaccines against *A. hydrophila*. This is a technical difficulty, which may be overcome if it is possible to identify common antigens which induce protection. Beyond this, the concerns and limitations are the economic factors. Infections with *A. hydrophila* are most significant in warmwater fisheries, and often in countries where the cost of a commercially developed, licensed vaccine would be prohibitive. In salmonid aquaculture and other operations where vaccines might be cost-effective, improvements in husbandry and environmental conditions are likely to have as much positive impact on the health of these fish as application of a vaccine against *A. hydrophila*. To some extent, vaccine usefulness would be limited to marginal situations. Unfortunately, these are also the situations in which parasitic infections or the use of antibiotics may alter immune responsiveness to vaccines.

Ruangapan *et al.*[4] suggested that immersion vaccination is not a feasible method of vaccine application for *A. hydrophila*, although this contradicts experience with other bacterial pathogens (see comments of Lamers *et al.*[42]

above). Injection immunization would add significantly to the cost of vaccination. In addition, injections of preparations containing adjuvant, which appears to enhance the protective effect, can themselves be a potential source of *A. hydrophila* infection.[44]

OVERCOMING LIMITATIONS

If labour costs are relatively low, or in special circumstances such as broodstock fish being held for long periods, injection vaccination might be feasible. For pond culture or open water stocking, however, an immersion vaccine that can be easily applied to small fish with simple equipment and little technical skill would be desirable. Immersion is also attractive when the importance of route of vaccination and either booster or challenge is considered; constant exposure to potentially pathogenic strains of *A. hydrophila* in the water may provide an enhancement of the initial vaccination treatment. Presumably an orally administered vaccine would have applications, but the limited trials suggest this is not as effective as other routes, and may induce additional costs for repeated applications.

It is the initial technical problem which needs addressing. It is not clear whether a few strains of *A. hydrophila* are the most significant in diseases, or whether many environmental strains are potential pathogens. If a few antigens that provide sufficient cross-protection can be identified, then a standard type of bacterin could be developed. If the problem of heterogeneity of strains cannot be overcome in this way, alternative approaches to protection from motile aeromonad disease may be through enhancement of natural resistance mechanisms (mentioned above), some of which may be stimulated by adjuvants.

COMMERCIAL PROSPECTS

Initial indications are not optimistic for the development of a vaccine for *A. hydrophila*, yet there are some potential opportunities. In some warmwater aquaculture operations, there may be very little flexibility to alter management of the environmental conditions, particularly at specific times of the year. In the Philippines and in Thailand, outbreaks appear most severe in the December to February period, with *A. hydrophila* being an opportunistic infection of stressed fish or a secondary invader of parasite-injuries. A vaccine might reduce losses during this critical period. A second potential opportunity is the use of a vaccine as an inexpensive replacement for antibiotics in intensive culture or holding. Eels in shipping containers are susceptible to losses due to *A. hydrophila*, yet antibiotic treatment delays the time at which they can be sold for human consumption in many countries. As vaccination appears to be effective in eels, there is a real possibility for a commercial vaccine, even an injection vaccine, to be cost-effective. Similarly, an immersion vaccination would be of value in treating baitfish, where antibiotic administration presents the opportunity for development of resistance in the bacteria associated with the fish. A vaccine to *A. hydrophila* would be most useful where the options to improve management or environment are limited, when outbreaks are regular occurrences, and when antibiotic use is restricted by cost or regulation.

CONCLUSIONS

A. hydrophila infections have received relatively little attention from vaccine producers. Contributing to this have been difficulties with an appropriate challenge and perceptions about the serological heterogeneity of this group of bacteria. The ability of warmwater aquaculture operations to pay the commercial costs of a vaccine is also in question. However, if aeromonads share very similar virulence factors, work on furunculosis vaccines may allow bacterins for *A. hydrophila* to be developed fairly rapidly.

REFERENCES

1 Hazen, T. C., Esch, G. W., Glassman, A. B. and Gibbons, J. W. (1978). Relationship of season, thermal loading and red-sore disease with various haematological parameters in *Micropterus salmoides*. *J. Fish Biol.*, **12**, 491–498.

2 Elliot, D. and Shotts, E. Jr (1980). Aetiology of an ulcerative disease in goldfish, *Carassius auratus* L.: Experimental induction of the disease. *J. Fish Dis.*, **3**, 687–693.

3 Amin, N. E., Abdallah, I. S., Elallawy, T. and Ahmed, S. M. (1985). Motile *Aeromonas* septicaemia among *Tilapia nilotica* (*Sarotherodon niloticus*) in Upper Egypt. *Fish Pathol.*, **20**, 93–97.

4 Ruangapan, L., Kitao, T. and Yoshida, T. (1986). Protective efficacy of *Aeromonas hydrophila* vaccines in Nile tilapia. *Veter. Immunol. Immunopathol.*, **12**, 345–350.

5 Newman, S. (1983). *Aeromonas hydrophila*: A review with emphasis on its role in fish disease. In: *Antigens of Fish Pathogens*, ed. D. P. Anderson, M. Dorson and Ph. Dubourget, Collection Fondation Marcel Merieux, pp. 87–117.

6 Miyazaki, T. and Jo, Y. (1985). A histopathological study of motile aeromonad disease in ayu *Plecoglossus altivelis*. *Fish Pathol.*, **20**, 55–60.

7 Groberg, W. J., McCoy, R. H., Pilcher, K. S. and Fryer, J. L. (1978). Relation of water temperature to infections of coho salmon (*Oncorhynchus kisutch*), chinook salmon (*O. tshawytscha*), and steelhead trout (*Salmo gairdneri*) with *Aeromonas salmonicida* and *A. hydrophila*. *J. Fish. Res. Board Can.*, **35**, 1–7.

8 Nieto, T. P., Corcobado, M. J. R., Toranzo, A. E. and Barja, J. L. (1985). Relation of water temperature to infection of *Salmo gairdneri* with motile *Aeromonas*. *Fish Pathol.*, **20**, 99–105.

9 Kaper, J. B., Lockman, H. and Colwell, R. R. (1981). *Aeromonas hydrophila:* Ecology and toxigenicity of isolates from an estuary. *J. Applied Bacteriol,* **50**, 359–377.

10 Kirov, S. M., Rees, B., Wellock, R. C., Goldsmid, J. M. and Van Galen, A. D. (1986). Virulence characteristics of *Aeromonas* spp. in relation to source and biotype. *J. Clin. Microbiol.*, **24**, 827–834.

11 Olah, J., Sharangi, N. and Datta, N. C. (1986). City sewage fish ponds in Hungary and India. *Aquaculture,* **54**, 129–134.

12 De Figueirredo, J. and Plumb, J. A. (1977). Virulence of different isolates of *Aeromonas hydrophila* in channel catfish. *Aquaculture* **11**, 349–354.

13 Mittal, K. R., Lalonde, G., Leblanc, D., Olivier, G. and Lallier, R. (1980). *Aeromonas hydrophila* in rainbow trout: relation between virulence and surface characteristics. *Can. J. Microbiol.*, **26**, 1501–1503.

14 Wakayabashi, H., Kanai, K., Hsu, T. C. and Egusa, S. (1981). Pathogenic activities of *Aeromonas hydrophila* biovar hydrophila (Chester) Popoff and Vernon, 1976 to fishes. *Fish Pathol.*, **15**, 319–325.

15 Leblanc, D., Mittal, K., Olivier, G. and Lallier, R. (1981). Serogrouping of motile *Aeromonas* species isolated from healthy and moribund fish. *Appl. Environ. Microbiol.*, **42**, 56–60.

16 Popoff, M. and Lallier, R. (1984). Biochemical and serological characteristics of *Aeromonas*. In: *Methods in Microbiology, Vol. 16,* ed. T. Bergan, London: Academic Press, pp. 127–145.

17 Dooley, J. G. S., Lallier, R., Shaw, D. H. and Trust, T. J. (1985). Electrophoretic and immunochemical analyses of the lipopolysaccharides from various strains of *Aeromonas hydrophila*. *J. Bacteriol.*, **164**, 263–269.

18 Dooley, J. S. G., Lallier, R. and Trust, T. J. (1986). Surface antigens of virulent strains of *Aeromonas hydrophila*. *Veter. Immunol. Immunopath.*, **12**, 339–344.

19 Boulanger, Y., Lallier, R. and Cousineau, G. (1977). Isolation of enterotoxigenic *Aeromonas* from fish. *Can. J. Microbiol.*, **23**, 1161–1164.

20 Hsu, T. C., Shotts, E. B. and Waltman, W. D. (1985). Action of *Aeromonas hydrophila* complex on carbohydrate substrates. *Fish Pathol.*, **20**, 23–35.

21 Ljungh, A. and Wadstrom, T. (1982). *Aeromonas* toxins. *Pharmac. Ther.*, **15**, 339–354.

22 Thune, R. L., Johnson, M. C., Graham, T. E. and Amborski, R. L. (1986). *Aeromonas hydrophila* β-haemolysin: purification and examination of its role in virulence in 0-group channel catfish, *Ictalurus punctatus* (Rafinesque). *J. Fish Dis.*, **9**, 55–61.

23 Leung, K-Y. and Stevenson, R. M. W. (1988). Characteristics and distribution of extracellular proteases from *Aeromonas hydrophila*. *J. Gen. Microbiol.*, **134**, 151–160.

24 Hsu, T. C., Waltman, W. D. and Shotts, E. B. (1981). Correlation of extracellular enzymatic activity and biochemical characteristics with regards to virulence of *Aeromonas hydrophila*. *Develop. Biol. Stand.*, **49**, 101–111.

25 Allan, B. J. and Stevenson, R. M. W. (1981). Extracellular virulence factors of *Aeromonas hydrophila* in fish infections. *Can. J. Microbiol.*, **27**, 1114–1122.

26 Thune, R. L., Graham, T. E., Riddle, L. M. and Amborski, R. L. (1982). Extracellular products and endotoxin from *Aeromonas hydrophila:* effects on age-0 catfish. *Trans. Amer. Fish. Soc.* **111**, 404–408.

27 Lallier, R., Bernard, F. and Lalonde, G. (1984). Difference in the extracellular products of two strains of *Aeromonas hydrophila* virulent and weakly virulent for fish. *Can. J. Microbiol.*, **30**, 900–904.
28 Thune, R. L., Graham, T. E., Riddle, L. M. and Amborski, R. L. (1982). Extracellular proteases from *Aeromonas hydrophila*: Partial purification and effects on age-0 channel catfish. *Trans. Amer. Fish. Soc.*, **111**, 749–754.
29 Leung, K-Y. (1987). The role of proteases of *Aeromonas hydrophila* in infections of rainbow trout. Ph.D thesis, University of Guelph, Ontario, Canada.
30 AQUIGRUP (1980). Trial vaccination of rainbow trout against *Aeromonas liquefaciens*. In: *Fish Diseases; Third COPRAQ-Session*, ed. W. Ahne, Berlin: Springer-Verlag, pp. 206–211.
31 Aoki, T., Egusa, S., Ogata, Y. and Watanabe, T. (1971). Detection of resistance factors in fish pathogen *Aeromonas liquefaciens*. *J. Gen. Microbiol.*, **65**, 343–349.
32 Rahim, Z., Sanyal, S. C., Aziz, K. M. S., Huq, M. I. and Chowdhury, A. A. (1984). Isolation of enterotoxigenic, haemolytic, and antibiotic-resistant *Aeromonas hydrophila* strains from infected fish in Bangladesh. *Appl. Envir. Microbiol.*, **481**, 865–867.
33 Shotts, E. B., Jr, Vanderwork, V. L. and Campbell, L. M. (1976). Occurrence of R factors associated with *Aeromonas hydrophila* isolates from aquarium fish and waters. *J. Fish. Res. Board Can.*, **33** 736–740.
34 Wu, J-L., Lin, H-M., Jan, L., Hsu, Y-L. and Chang, L-H. (1981). Biological control of fish bacterial pathogen, *Aeromonas hydrophila*, by bacteriophage AH1. *Fish Pathol.*, **15**, 271–276.
35 Plumb, J. A. (1984). Immunization of warm water fish against five important pathogens. In: *Symposium on Fish Vaccination*, ed. P. de Kinkelin, Paris: Office International des Epizooties, pp. 199–222.
36 Lamers, C. H. J. and Van Muiswinkel, W. B. (1986). Natural and acquired agglutinins to *Aeromonas hydrophila* in carp (*Cyprinus carpio*). *Can. J. Fish. Aquat. Sci.*, **43**, 619–624.
37 Davis, J. F. and Hayasaka, S. S. (1984). The enhancement of resistance of the American eel *Anguillarum rostrata* to the pathogenic bacterium *Aeromonas hydrophila* by extract of the tunicate *Ecteinascidia turbidinata*. *J. Fish Dis.* **7,** 311–316.
38 Lamers, C. H. J. (1985). The Reaction of the Immune System of Fish to Vaccination. Thesis, Agricultural University of Wageningen, The Netherlands.
39 Post, G. (1966). Response of rainbow trout (*Salmo gairdneri*) to antigens of *Aeromonas hydrophila*. *J. Fish. Res. Board Can.*, **23**, 1487–1494.
40 Khalifa, K. A. and Post, G. (1976). Immune response of advanced rainbow trout fry to *Aeromonas liquefaciens*. *Prog. Fish-Cult.*, **38**, 66–68.
41 Lamers, C. H. J., De Haas, M. J. M. and Van Muiswinkel, W. B. (1985). Humoral response and memory formation in carp *Cyprinus carpio* after injection of *Aeromonas hydrophila* bacterin. *Develop. Comp. Immunol.*, **9**, 65–76.
42 Lamers, C. J., De Haas, M. J. M. and Van Muiswinkel, W. B. (1985). The reaction of the immune system of fish to vaccination. Development of immunological memory in carp *cyprinus carpio* following direct immersion in *Aeromonas hydrophila* bacterin. *J. Fish Dis.*, **8**, 253–262.
43 Lamers, C. H. J. and De Haas, M. J. M. (1985). Antigen localization in the lymphoid organs of carp (*Cyprinus carpio*). *Cell Tissue Res.*, **242**, 491–498.
44 Munn, C. B. and Trust, T. J. (1983). Infection of rainbow trout *Salmo gairdneri* by opportunistic pathogens following sub-cutaneous injection of Freund's adjuvant. *Develop. Comp. Immunol.*, **7**, 193–194.

10

Vaccination against Bacterial Kidney Disease

A. L. S. Munro and D. W. Bruno

INTRODUCTION

Bacterial kidney disease (BKD) caused by *Renibacterium salmoninarum* (RS)[1] may cause serious mortality if introduced into intensively cultured populations of salmonid fish. Antimicrobial compounds are often less than efficacious in treatment and no vaccine is available. Unless infected populations are eradicated and the re-entry of the infectious agent prevented, disease recurrence is probable. As these criteria are often difficult or impossible to meet and because movements of cultured fish are spreading infection, this disease could easily become one of the most significant limiting commercial production of some salmonid species.

The purpose of this review is to consider the prospects for developing a BKD vaccine and what, if any, limitations it might have. Broader aspects of this disease have been reviewed by Fryer and Sanders.[2]

DISTRIBUTION AND SIGNIFICANCE OF BKD

The bacterium is endemic in wild anadromous salmonid populations on both coasts of north America.[2] It has been found in wild Atlantic salmon, *Salmo salar* L., and sea trout, *S. trutta* L., in the UK.[3] It is probably more widespread in wild salmonids in Europe but because of low prevalence and limited study its true distribution is unknown. Because the bacterium is egg- and fish-transmitted culturists have been instrumental in the widespread distribution of this obligate fish pathogen in Europe, as well as America, Alaska, Canada, Iceland, Chile and Japan.

Cultured fish are at risk of infection from surface waters carrying infected wild fish and from the introduction of other cultured and infected fish or ova. The disease is also troublesome in culture systems with high fish densities and or low water flows and in cage farms in fresh and sea water where young juveniles are placed alongside older survivors of previous infections. There is also increasing evidence that infected juvenile salmonids

FISH VACCINATION
ISBN 0-12-237485-1

destined for release to the wild for fishery enhancement subsequently develop the disease at sea and may continue to die there.[4,5]

AETIOLOGICAL AGENT

Morphology

RS is a Gram-positive, short rod (0.8–1.0 × 0.3–0.5 μm) which is non-motile, asporogenous, non-acid fast and not encapsulated.

Habitat

RS is an obligate pathogen for salmonid fish, found both intracellularly and extracellularly. However, in a recent survey (Eunice Lam, pers. comm.) RS was isolated from Pacific herring, *Clupea harengus pallasi* L., in cages which were holding coho salmon, *Oncorhynchus kisutch* (Walbaum) suffering clinical BKD and Hicks *et al.*[6] have experimentally infected fathead minnows, *Pimephales promelas* (Rafinesque) and the common shiner, *Notropis cornutus* (Mitchill).

Culture

Currently two media are in regular use, the commercial preparation Mueller-Hinton, and the serum based medium, KDM-2, developed by Evelyn.[7] In addition a semi-defined and a selective medium have been described by Embley *et al.*[8] and Austin *et al.*[9] respectively. All media require the addition of 0.1 per cent L-cysteine hydrochloride. Growth is slow, often requiring several weeks for primary isolations from fish tissues at 15°C. On Mueller-Hinton medium with cysteine hydrochloride virulent strains are sticky, smooth, slightly raised, creamy yellow and shiny.[10]

Biochemical properties

Isolates of RS from natural disease outbreaks from the UK, Canada, USA and Japan show considerable uniformity in their properties. Positive reactions include β-haemolysis, gelatin liquefaction, catalase and DNase production whereas negative reactions include cytochrome oxidase, methyl red, indole production and acid production from sugars.[11] The enzyme profile obtained using the API-ZYM system (API Laboratory Products),[9] provides a unique profile amongst bacterial fish pathogens and on its own is sufficient to distinguish RS from other bacteria.[12,13]

Chemical composition and antigenicity

Fatty acid composition. Embley *et al.*[14] studying the fatty acid profiles of 21 strains of RS showed they were composed almost exclusively of methyl-branched fatty acids. All strains contained several unsaturated menaquinones, and very characteristic polar lipid patterns consisting of diphosphatidylglycerol, two major and six or seven minor glycolipids and two unidentified phospholipid components. No mycolic acids, phosphatidylinositol or related dimannosides, or diaminobutyric acid were found, confirming other single strain studies e.g. see Goodfellow *et al.*[13]

DNA content. The guanine plus cytosine (G + C) content of the bacteria averaged 53 mol per cent.[1]

Cell wall composition. Analysing the composition of the cell wall of 13 strains of RS, Fiedler and Draxl[15] found a marked similarity in both peptidoglycan and the covalently linked cell wall polysaccharide confirming earlier work of Kusser and Fiedler.[16] Of particular interest is the cell wall polysaccharide which amounted to more than 60 per cent of the dry weight of the cell walls and contained galactose as the major sugar with lesser amounts of rhamnose, *N*-acetylglucosamine, and *N*-acetylfucosamine. The purified cell wall polysaccharide was shown to be antigenic in rabbits.

Extracellular products. Kaattari *et al.*[17] have shown that an RS strain produces three extracellular soluble proteins (of 60, 34 and 26 kd) as resolved by SDS polyacrylamide gel electrophoresis. Turaga *et al.*[18] reported that, collectively, these ECP proteins in native form suppress *in vivo* antibody responses of lymphocytes. Additionally, in experimentally infected fish there was an association of decreasing haematocrit with increasing serum levels of soluble antigen supporting the conclusions of Bruno and Munro[19] that RS was producing factors causing the accumulation of erythrocytes in the spleen. Other reports[20,21] have demonstrated the presence of a heat stable antigen of RS in body fluids and tissues of infected fish.

Antigenicity of RS strains. Studies prior to and including Getchell *et al.*[22] report on the uniform antigenicity of isolates of RS using polyclonal antibodies prepared against whole cell preparations. However, using monoclonal antibodies Arakawa *et al.*[23] have demonstrated that antigenic differences exist between isolates.

PATHOLOGY

BKD is a chronic, often fatal disease causing both intra- and extracellular infection and arises from skin, eye and other possible, as yet unidentified, primary sites of infection. Subsequently, RS may localize in the kidney from which infection rapidly becomes systemic. Both gross and histological lesions of moribund salmonids have been well described e.g. Bruno.[24] However, the granulomatous lesions so often a characteristic of the disease will be discussed further because of their relevance to host protective responses. This lesion is a chronic, diffuse, inflammatory condition involving one or more of the soft tissues. It may be especially obvious when the serosal lining of organs e.g. kidney, liver, spleen or swim bladder is affected showing as a thickened opaque membrane. At the cellular level, fibroblasts often predominate in random proliferation with numerous monocytes, macrophages and lymphocytes attempting to clear bacteria, intracellularly infected host cells and necrotic debris.

Pathogenesis

Further information on the development of the disease should be considered before reaching conclusions about the adequacy of the host response in containing the bacterium. At least three aspects are important, host species and strain susceptibility, the influence of temperature and evidence (including the nature) of recovery. In addition, information on how the bacterium escapes or defeats the host defences may be important in the context of developing a vaccine. The intracellular location of RS is mostly in monocytes and macrophages although other weakly phagocytic cells also harbour the bacterium.[24] A correlation between increased cell surface hydrophobicity, auto-agglutination and virulence of RS strains may increase their rate of attachment and uptake into the host cells and also provide protection from the action of lysosomal enzymes within the phagosome.[10] Within the genus *Oncorhynchus*, Sanders *et al.*[25] and Bell *et al.*[26] consider pink salmon, *O. gorbushca* (Walbaum) the most susceptible followed by sockeye, *O. nerka* (Walbaum), chinook, *O. tshawytscha* (Walbaum) and perhaps the chum, *O. keta* (Walbaum) with coho, *O. kisutch* the least susceptible. Information on the genus *Salmo* is limited but in the authors' experience European strains of the rainbow trout, *S. gairdneri* (Richardson) and the Atlantic salmon have a modest resistance. Using the comparative data in Sanders *et al.*[25] all species in the genus *Oncorhynchus* are more susceptible to BKD than *Salmo* species with the steelhead trout, a strain of *S gairdneri*, amongst the more susceptible. In coho salmon, a transferrin genotype has been shown to be correlated with some limited degree of resistance to BKD.[27]

Natural outbreaks of BKD are strongly correlated with season occurring from late autumn to early spring.[28] As temperatures increase from spring to summer the prevalence of diseased fish, as a broad generalization, decreases and evidence of BKD may disappear entirely.[28] Bruno and Munro[29] studied the fate of a commercially reared population of post yearling parr which suffered a cumulative BKD mortality from December to smolting in May of 18 per cent when the fish were transferred to sea water. By August a further cumulative mortality of 16 per cent attributable to BKD occurred after which mortality stopped. Thereafter, RS could not be detected in any fish sampled by Gram and indirect FAT stain up to 69 weeks post seawater transfer. Sampling showed 39 per cent of the smolts were infected as measured by Gram stain at transfer to sea water and none had measurable levels of agglutinating antibody. In June, antibody was detected in some fish reaching a maximum in all sampled fish by September, 19 weeks post seawater transfer, and thereafter began to decline between December and April, weeks 31–49.[30] The evidence of agglutinin response in all fish indicates they had been infected at some time and the survivors effectively mounted a response eliminating infection and in some cases clinical disease. In post smolting fish cortisol levels will diminish and the fish would experience increasing seawater temperatures (9–14°C), both factors likely to assist in helping the fish mount effective immune responses. Unpublished histological observations (DWB) from this population clearly show the resolution of granulomatous lesions, showing all the stages from walling off, loss of Gram-positive staining material inside the walled-off area, and final dissolution of this lesion, which is interpreted as evidence of cell mediated immunity (CMI) resolving the disease and all evidence of the invading bacterium.

Experimental infection over a temperature range 4–20°C using coho salmon and steelhead trout showed mortality was maximal (78–100 per cent) in the range 7–12°C and as the temperature increased above 12°C mortality declined progressively to 8–14 per cent at 20°C; with sockeye salmon mortality reached 100 per cent at all temperatures.[25]

Cvitanich[31] has reported 'bar forms' of the RS bacterium, their characteristic appearance visualized by the direct fluorescent antibody technique (FAT). 'Bar forms', reputedly dead or dying bacteria, are reported in all species of the Salmonidae. Quantitative assessment of their numbers is reported to allow evaluation of disease severity and in consequence the percentage of the population which is terminally affected. The more intensely staining 'bar' dividing the cell into two halves may represent a host mediated mechanism interfering with bacterial cell division. Where 'bar forms' originate e.g. extracellularly or intracellularly, is uncertain. Collectively these reports suggest recovery is possible in most salmonids and that temperature is of critical importance. Clearance of bacteria from intracellular locations suggests that CMI is likely to assume greater importance than humoral immunity and may also explain why warmer temperatures are important.[32]

EXPERIMENTAL IMMUNOLOGY

Evelyn[33] first showed with 1–3 year old sockeye salmon held in fresh water at 12–15°C that an i.p. injection of killed cells of RS (450 μg/g of fish) in a mineral oil : Arlacel adjuvant produced agglutinating antibodies. In 30 days 1/10 fish responded, in 60 days 5/10 fish and in 90 days 8/10 fish, indicating a very slow response. In a separate experiment using Freund's complete adjuvant (FCA) and 200 μg antigen/g of fish, a response was still detectable after 16 months and when fish were given a secondary antigenic stimulation at 13 months their response 3 months later was far greater than the primary at 3 months indicating a strong anamnestic response. Evelyn *et al.*[34] evaluated various vaccine preparations and methods of their presentation in sockeye and coho salmon. The vaccines tested were all derived from formalin-killed RS cultures and included cell associated and extracellular antigens, whole cultures, intact cells and various crude cell fractions. The vaccines were administered i.p., with and without FCA, or by feeding, spraying, or the two step hyperosmotic infiltration method. The efficacy of the vaccines was determined by comparing mortalities in vaccinated and unvaccinated fish following natural and experimental (injected) challenge with the live pathogen. All vaccinated fish showed no evidence of protection.

Paterson *et al.*[35] investigated the immune response of under-yearling Atlantic salmon parr of 0.3–0.6 g, vaccinated by i.p. injection with 2.5×10^7 heat killed RS cells either with or without FCA, or bathed in 10^7 RS cells/ml, heat killed and ultrasonically disrupted, and with fish held at temperatures in excess of 10°C. Parr injected i.p. with RS and FCA demonstrated a strong agglutinin response first detectable at 5 weeks, peaking at 24 weeks but still high at 69 weeks and diminishing thereafter to nil at 90 weeks. Parr injected i.p. with RS alone gave no response and immersion vaccinated parr gave small but detectable responses over the same period. No agglutinins were detected in intestinal mucus whereas cutaneous mucus had small titres in fish vaccinated i.p. with RS/FCA. None of these vaccinated populations demonstrated protection to natural challenge 57–98 weeks later as measured by numbers of lesions or by prevalence of infection using indirect FAT. Fish vaccinated with RS/FCA showed a higher rate of RS injection than the other groups.

In the same report Paterson *et al.*[35] record the results of injected RS with FCA in post-yearling parr. Up to 52 weeks, the longest period tested, the parr showed a strong agglutinin response. After vaccination the fish were exposed to natural challenge. In three separate trials lesion prevalence was reduced to almost nil in vaccinated fish (20–30 per cent in controls) but infection rates as judged by FAT were the same in control and vaccinated fish. This latter experiment indicates a marked improvement in the health of the post-yearlings at the time of testing, about 40–46 weeks after vaccination. However, at that point the parr, now smolts, were ready for release to sea and another 1.5–2.5 years of life during which time it remains unknown if the vaccine offered protection from the infection which many were carrying. The failure of the vaccine in the under-yearlings was attributed to the long period between vaccination and challenge, about 1.5 years, although it must also be noted that with current knowledge vaccination is considered less effective in fish smaller than 2–4 g in size. The higher prevalence of lesions in the vaccinated fish is also suggestive of vaccination at too early an age.

McCarthy *et al.* (1984) have studied the efficacy of various RS immunizing preparations without adjuvant in rainbow trout. RS was grown in KDM-2 plus serum, and the best vaccination results were obtained using pH lysed cells in whole culture medium then formalized. Trout of 10–15 g were vaccinated by i.p. injection, by direct immersion in bacterin for 2 min or by the two step hyperosmotic infiltration procedure using 5.6 per cent NaCl. Fish were challenged after holding at 11°C for 34 days by i.p. injection or immersion using undiluted broth cultures of homologous pathogen. In preliminary challenge trials using non-vaccinated fish, i.p. challenge resulted in 75 per cent mortality at an average time of 30 days whereas bathing caused 14 per cent and continuing mortality by 40 days illustrating the length of time necessary to conduct vaccine trials with a chronic disease such as BKD. In subsequent tests of the immunizing methods i.p. challenge was used and the results showed that immersion and hyperosmotic infiltration gave no protection. In fish vaccinated by i.p. injection the best vaccine preparation gave 4 per cent infection (sum of mortalities + infected fish at 40 days post challenge) against 90 per cent in controls.

CONCLUSIONS

Atlantic salmon suffering a natural BKD outbreak are capable of recovering and apparently eliminating the infection.[29] Recovery occurred after transfer to sea water and an increase in ambient temperature. It is also possible that smolts are immunologically suppressed.[37] Alone these observations do not distinguish between the possibility that a non-specific intracellular killing mechanism is activated by increasing temperature or that the CMI response was similarly activated. However, the parallel observations on lesion recovery and the nature of the cells associated with it do suggest a strong CMI involvement.

Although the foregoing examples are not numerous it is concluded that systemic RS infection may confer protection from overt disease in salmonids which may even result in the elimination of the infection. This evidence is stronger for *Salmo* than for *Oncorhynchus*. As yet there is no information on the immunogenicity of defined BKD antigens in fish, nor the nature of antigens which stimulate a protective immune response.

The results of experimental vaccination indicate in all species a significant but slow to develop, non-protective agglutinating antibody response. The major conclusion from this should be that the immunizing preparations are inadequate perhaps because they contain the wrong antigens, insufficient of those necessary to confer protection or perhaps the important antigens are too weakly immunogenic or responses to them suppressed by other antigens. In this respect an evaluation and role of immuno-modulators could be considered. Hastings and Ellis[38] comparing the responses of the rabbit and rainbow trout when immunized with the 30 or so detectable *Aeromonas salmonicida* extracellular proteins showed that the rabbit produced antibodies to at least 15, whereas antibodies to only five were detected in the trout. Rabbit antibodies to *A. salmonicida* have also been shown by Olivier *et al.*[39] to protect fish by passive immunization whereas the same antigenic preparations did not raise protective antibodies in fish.

Immunosuppressive and other possible *in vivo* effects of RS ECP's remain to be more fully evaluated but it appears they may have significant activity on fish immune function. The action of some individual products may be of critical importance in understanding both the failure of RS vaccines so far and how to make a successful one. They may also hold answers to why some salmonids are more susceptible to BKD. The separation and characterization of activity of individual RS ECP's is therefore likely to be important in the development of any successful vaccine.

Those experimental vaccines showing some protection failed to eliminate infection in all fish yet recovery from a natural epidemic apparently did. Of course the short time scale of the experimental regimes may not reflect the final outcome but it does illustrate that work on a BKD vaccine requires much more time for testing than vaccines for other fish diseases of an acute nature. The long life span of sea cultured and sea ranched salmonids also indicates that a vaccine should offer protection for not less than a year and probably much longer. The strong implication of the importance of cell mediated immunity for adequate protection, its dependence on a minimum temperature for an adequate response and an *in vivo* assay system for measuring protection are other important considerations in any vaccination strategy.

REFERENCES

1 Sanders, J. E. and Fryer, J. L. (1980). *Renibacterium salmoninarum* gen. nov., sp. nov., the causative agent of bacterial kidney disease in salmonid fishes. *Int. J. Syst. Bact.*, **30**, 496–502.

2 Fryer, J. L. and Sanders, J. E. (1981). Bacterial kidney disease of salmonid fish. *Ann. Rev. Microbiol.*, **35**, 273–298.

3 Smith, I. W. (1964). The occurrence and pathology of Dee disease. *Freshwat. Salmon Fish. Res.* **34**, 1–12.

4 Ellis, R. W., Novotny, A. J. and Harrell, L. W. (1978). Case report of kidney disease in a wild chinook salmon, *Oncorhynchus tshawyscha,* in the sea. *J. Wild. Dis.*, **14**, 120–123.

5 Banner, C. R., Long, J. J. and Rohovec, J. S. (1986). Occurrence of salmonid fish infected with *Renibacterium salmoninarum* in the Pacific Ocean. *J. Fish Dis.*, **9**, 273–275.

6 Hicks, B. D., Daly, J. G. and Ostland, V. E. (1986). *Experimental Infection of Minnows with the Bacterial Kidney Disease Bacterium.* Abstract from a 1986 Aquaculture Association of Canada meeting, University of Guelph, Canada 3–7 June 1986.

7 Evelyn, T. P. T. (1977). An improved growth medium for the kidney disease bacterium and some notes on using the medium. *Bull. Off. int. Epiz.*, **87**, 511–513.

8 Embley, T. M., Goodfellow, M. and Austin, B. (1982). A semi-defined medium for *Renibacterium salmoninarum. FEMS Microbiol. Lett.*, **14**, 299–301.

9 Austin, B., Embley, T. M. and Goodfellow, M. (1983). Selective isolation of *Renibacterium salmoninarum. FEMS Microbiol. Lett.*, **17**, 111–114.

10 Bruno, D. W. (1988). The relationship between auto-agglutination, cell surface hydrophobicity and virulence of the fish pathogen, *Renibacterium salmoninarum. FEMS Microbiol. Lett.*, **51**, 135–140.

11 Bruno, D. W. and Munro, A. L. S. (1986). Uniformity in the biochemical properties of *Renibacterium salmoninarum* isolates obtained from several sources. *FEMS Microbiol. Lett.*, **33**, 247–250.

12 Collins, M. D. (1982). Lipid composition of *Renibacterium salmoninarum* (Sanders and Fryer). *FEMS Microbiol. Lett.*, **13**, 295–297.

13 Goodfellow, M., Embley, T. M. and Austin, B. (1985). Numerical taxonomy and embedded description of *Renibacterium salmoninarum. J. Gen. Microbiol.*, **131**, 2739–2752.

14 Embley, T. M., Goodfellow, M., Minnikin, D. E. and Austin, B. (1983). Fatty acids, isoprenoid quinone and polar lipid composition in the classification of *Renibacterium salmoninarum. J. Appl. Bacteriol.*, **55**, 31–37.

15 Fiedler, F. and Draxl, R. (1986). Biochemical and immunological properties of the cell surface of *Renibacterium salmoninarum. J. Bacteriol.*, **168**, 799–804.

16 Kusser, W. and Fiedler, F. (1983). Murein type and polysaccharide composition of cell walls from *Renibacterium salmoninarum. FEMS Microbiol. Lett.*, **20**, 391–394.

17 Kaattari, S., Getchell, R., Turaga, P. and Irwin, M. (1985). *Development of a Vaccine for Bacterial Kidney Disease in Salmon.* Annual report FY 1984, Bonnerville Power Administration, Portland, 55 pp.

18 Turaga, P., Wiens, G. and Kaattari, S. (1987). Bacterial kidney disease: the potential role of soluble protein antigen(s). *J. Fish Biol.*, **31** (Supplement A), 191–194.

19 Bruno, D. W. and Munro, A. L. S. (1986). Haematological assessment of rainbow trout, *Salmo gairdneri* Richardson, and Atlantic salmon, *Salmo salar* L., infected with *Renibacterium salmoninarum. J. Fish Dis.*, **9**, 195–204.

20 Kimura, T., Ezura, Y., Tajima, K. and Yoshimizu, M. (1978). Serological diagnosis of bacterial kidney disease of salmonid (BKD): Immunodiffusion test by heat stable antigen extracted from infected kidney. *Fish Pathol.*, **13**, 103–108.

21 Pascho, R. J. and Mulcahy, D. (1987). Enzyme-linked immunosorbent assay for a soluble antigen of *Renibacterium salmoninarum*, the causative agent of salmonid bacterial kidney disease. *Can. J. Fish. Aquat. Sci.*, **44**, 183–191.
22 Getchell, R. G., Rohovec, J. S. and Fryer, J. L. (1985). Comparison of *Renibacterium salmoninarum* isolates by antigenic analysis. *Fish Pathol.*, **20**, 149–159.
23 Arakawa, C. K., Sanders, J. E. and Fryer, J. L. (1987). Production of monoclonal antibodies against *Renibacterium salmoninarum*. *J. Fish Dis.*, **10**, 249–253.
24 Bruno, D. W. (1986). Histopathology of bacterial kidney disease in laboratory infected rainbow trout, *Salmo gairdneri*, Richardson, and Atlantic salmon, *Salmo salar* L., with reference to naturally infected fish. *J. Fish Dis.*, **9**, 523–537.
25 Sanders, J. E., Pilcher, K. S. and Fryer, J. L. (1978). Relation of water temperature to bacterial kidney disease in coho salmon (*Oncorhynchus kisutch*), sockeye salmon (*O. nerka*), and steelhead trout (*Salmo gairdneri*). *J. Fish. Res. Board Can.*, **35**, 8–11.
26 Bell, G. R., Higgs, D. A. and Traxler, G. S. (1984). The effect of dietary ascorbate, zinc and manganese on the development of experimentally induced bacterial kidney disease in sockeye salmon (*Oncorhynchus nerka*). *Aquaculture*, **36**, 293–311.
27 Winter, G. W., Schreck, C. B. and McIntyre, J. D. (1980). Resistance of different stocks and transferrin genotypes of coho salmon, *Oncorhynchus kisutch*, and steelhead trout, *Salmo gairdneri*, to bacterial kidney disease and vibriosis. *Fish. Bull.*, **77**, 795–802.
28 Bruno, D. W. (1986). Scottish experience with bacterial kidney disease in farmed salmonids between 1976 and 1985. *Aquacult. Fish. Managmt.*, **17**, 185–190.
29 Bruno, D. W. and Munro, A. L. S. (1982). *First Experience of Bacterial Kidney Disease in the Sea Water Culture of Atlantic salmon* (Salmo salar). International Council for the Exploration of the Sea, ICES CM 1982/F:31, 4pp.
30 Bruno, D. W. (1987). Serum agglutinating titres against *Renibacterium salmoninarum* the causative agent of bacterial kidney disease, in rainbow trout, *Salmo gairdneri* Richardson, and Atlantic salmon, *Salmo salar* L. *J. Fish Biol.*, **30**, 327–334.
31 Cvitanich, J. (1987). Renibacterium salmoninarum '*Bar forms*': *Evidence of a Host Response to Bacterial Kidney Disease Infection*, Abstracts of Fish Immunology Symposium, Plymouth, England.
32 Ellis, A. E. (1988). Immunology of teleosts. In: *Fish Pathology*, 2nd edn, ed. R. J. Roberts, London: Baillière Tindall.
33 Evelyn, T. P. T. (1971). The agglutinin response in sockeye salmon vaccinated intraperitoneally with a heat-killed preparation of the bacterium responsible for salmonid kidney disease. *J. Wildl. Dis.*, **7**, 328–335.
34 Evelyn, T. P. T., Ketcheson, J. E. and Prosperi-Porta, L. (1984). *On the Feasibility of Vaccination as a Means of Controlling Bacterial Kidney Disease in Pacific Salmon*. Abstract of International Conference on the Biology of Pacific Salmon, 5–12 September 1984, Victoria/Agassiz, British Columbia.
35 Paterson, W. D., Desautels, D. and Weber, J. M. (1981). The immune response of Atlantic salmon, *Salmo salar* L., to the causative agent of bacterial kidney disease *Renibacterium salmoninarum*. *J. Fish Dis.*, **4**, 99–111.
36 McCarthy, D. H., Croy, T. R. and Amend, D. F. (1984). Immunisation of rainbow trout, *Salmo gairdneri* Richardson, against bacterial kidney disease: preliminary efficacy evaluation. *J. Fish Dis.*, **7**, 65–71.

37 Maule, A. G., Schreck, C. B. and Kaattari, S. L. (1987). Changes in the immune system of coho salmon (*Oncorhynchus kisutch*) during the parr-to-smolt transformation and after implantation of cortisol. *Can. J. Fish. Aquat. Sci.*, **44**, 161–166.
38 Hastings, T. S. and Ellis, A. E. (1988). The humoral immune response of rainbow trout and rabbits to *Aeromonas salmonicida* extracellular products. *J. Fish Dis.*, **11**, 147–160.
39 Olivier, G., Evelyn, T. P. T. and Lallier, R. (1985). Immunogenicity of vaccines from a virulent and an avirulent strain of *Aeromonas salmonicida*. *J. Fish Dis.*, **8**, 43–55.

11

Vaccination against *Edwardsiella tarda*

Fulvio Salati

The increased demands for higher production in aquaculture result in the appearance of previously undescribed diseases. Ewardsiellosis is a recently described bacterial disease of cultured fish caused by *Edwardsiella tarda*, Ewing and McWhorter, 1965, synonym *E. anguillimortifera*. The bacterium belongs to the family Enterobacteriaceae, is a Gram-negative, facultatively anaerobic, motile, peritrichously flagellated, non-encapsulated, non-spore-forming rod which grows from 15 to 42°C. Optimum growth conditions are 30°C, pH 5.5 to 9.0 and NaCl 0.5 to 4.0 per cent. *Ewardsiella tarda* was first reported as the causative agent of disease in the Japanese eel[1,2] and later in a variety of cultured fish of fresh and sea water such as goldfish, rainbow trout, chinook salmon, tilapia, mullet, crimson sea bream, red sea bream and flounder.[3–8] The microorganism has also been reported as a normal inhabitant of the intestine of reptiles,[9] of aquatic animals[10] including birds, alligators, turtles, frogs, mussels, brown bullhead and largemouth bass and as a possible source of human sporadic diarrhoea in the tropics.[11]

THE DISEASE PROBLEM

Edwardsiella tarda is distributed worldwide. However, edwardsiellosis is a problem principally in Africa, America and Asia and is an important disease of cultured eels in Japan and Taiwan. The losses attributed to *E. tarda* infection in cultured eel in Japan for the year 1984 are about 815 million yen, which corresponds to $\frac{1}{4}$ of the total losses. Thus, edwardsiellosis has been ranked as a disease of primary importance in eel culture. Generally, in aquaculture, edwardsiellosis occurs in spring and summer when the water temperature is higher than 24°C. In Japan, eel culture employs two methods: the Shizuoka method using large open ponds and mud bottom, and the Kochi method which uses medium size ponds, constructed of a concrete bottom, lined with vinyl sheets. This method is widely used and the water temperature can rise to 27°C or more in the hottest period of the year. The period of risk of edwardsiellosis for cultured eels begins when the glass eels

FISH VACCINATION
ISBN 0-12-237485-1

(postlarval stage) are introduced into the pond and start feeding. The period of risk continues with possible mortalities up to 90–95 per cent in the case of natural infection during the elver stage (a young eel), then the mortalities decrease to levels of 70–75 per cent in anguillettes (more than 40 g eel). Experimental infection by injection has demonstrated that elvers are more susceptible to edwardsiellosis than the anguillettes.[12]

The infectivity of *E. tarda* is not very high. However, the bacterium is thought to survive in eel ponds throughout the rearing period and causes disease when the physiological condition of the fish or the quality of the water deteriorates. For this reason it is very important to control the water conditions but practically it is not desirable to change the water temperature or the stocking density because high water temperature and high stocking density are the conditions *sine qua non* for good production in eel culture. *E. tarda* is commonly isolated from the intestine of the eel and often from the mud of the ponds.[13] It is difficult to eliminate the pathogen from the environment, as there is a variety of natural reservoirs including worms, reptiles and birds. Passive transmission by water, equipment, tools and food may occur. No vertical transmission of the disease is suspected.

E. tarda possesses pili on the cell surface and it is likely that these are the means by which the bacterium initially attaches to its host and establishes infection.[14] Two forms of infection have been reported in the eel:[15,16] nephritic and hepatic but when the disease is advanced it becomes systemic involving most organs (Fig. 11.1). Recently, the production of exotoxic substances by *E. tarda* has been reported.[17,18] A few strains produce haemolysins and all strains show dermatotoxic activity in the Japanese white rabbit. The pathogenesis of edwardsiellosis is not well understood. However, it is supposed that the penetration of the bacteria into the body of the eel is via cutaneous lesions or the intestine.[14]

Diagnosis of edwardsiellosis is based on the isolation and identification of the etiological agent and recently, the fluorescent antibody technique (FAT) and enzyme linked immunosorbent assay (ELISA) have been used for the rapid detection of the bacteria. The heterogeneity of *E. tarda* strains is enormous. There are strains that are only saprophytes and there are strains which are virulent and pathogenic e.g. the lethal dose of strain EF-1, isolated from an outbreak of disease in the eel, was $10^{6.8}$ cells/eel.[19] Little is known about the antigenic properties of *E. tarda*: 49 O-antigens, 37 H-antigens and 148 O–H combinations have been reported.[20]

In vitro, *E. tarda* shows sensitivity to various chemotherapeutants. These are, in order of decreasing sensitivity, oxolinic acid, trimethoprim, furazolidone, oxytetracycline, chloramphenicol and piromidic acid. In the field, diseased eels can be cured by oxolinic acid, oxytetracycline and piromidic acid. Treatment with sulphonamide will reduce the losses but improvements in hygiene, water quality and stocking density are also necessary. Nevertheless, the indiscriminate use of chemotherapeutants in aquaculture and the appearance of drug-resistant strains of the pathogen is problematical.

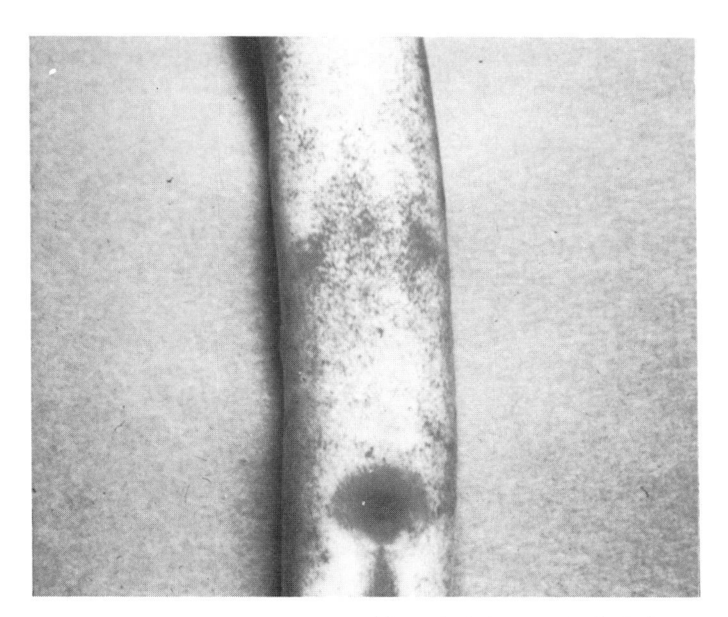

Fig. 11.1 Japanese eel, *Anguilla japonica*, with typical haemorrhagic lesions caused by *Edwardsiella tarda*. (Photo courtesy of Professor Dr R. Kusuda, 1974.)

THE IDEAL VACCINE

From the foregoing description of the disease problem, the ideal vaccine should:

1 Be delivered by immersion or orally since the risk of mortalities from edwardsiellosis is principally amongst elvers.
2 Provide protection throughout the susceptible stages of life i.e. up to and including anguillettes.
3 Be capable of inducing protective immunity at temperatures below 24°C (i.e. during winter and spring).
4 Protect against the different strains of the pathogen.
5 Be killed preparations or easily produced antigen extracts.

NATURE OF *E. TARDA* VACCINES

Basically, two types of vaccines have hitherto been used – bacterins and bacterial extracts. Bacterins as formalin killed cells were obtained by treating *E. tarda* cells with 0.5 per cent formalin; heat killed cells were obtained by heating the cells at 100°C for 30 min (Fig. 11.2). Bacterial extracts have been derived from culture supernatants or extracts of the cell wall. The latter preparations include lipopolysaccharide, extracted from the bacteria by the hot phenol–water method of Westphal and Jann,[24] crude LPS, purified LPS, polysaccharide and lipid A (Figs 11.3–7).

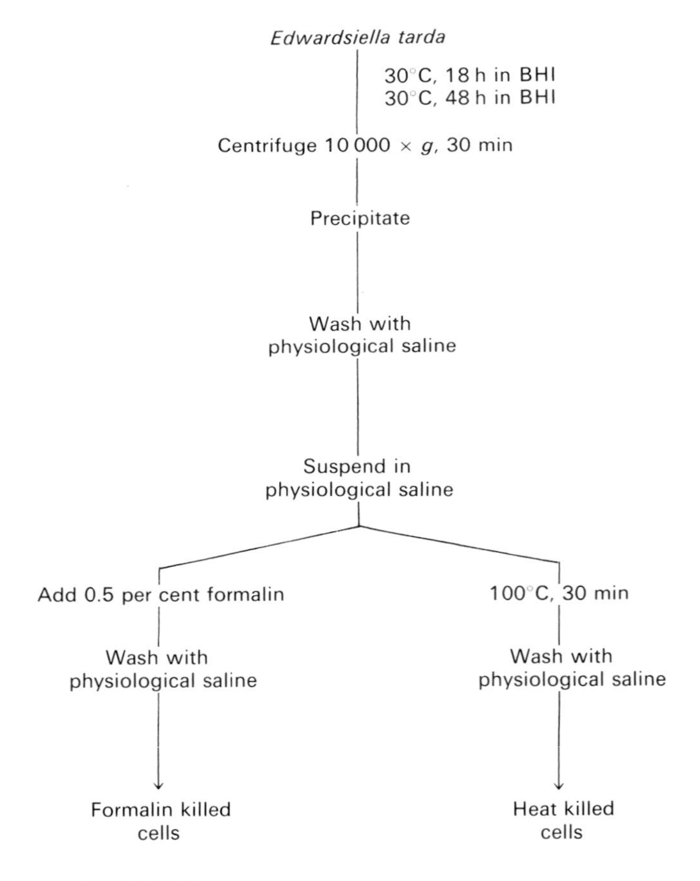

Fig. 11.2 Schematic drawing for preparation of bacterins.

EXPERIMENTAL VACCINATION STUDIES

To date only experimental studies have been performed. Initially the use of bacterins (see above) demonstrated that vaccination could provide a degree of protection but that this was insufficient for commercial exploitation. More recently, the search for more effective semi-purified antigens has resulted in greater optimism. While *E. tarda* infection is a problem for various species of cultured fish, most immunization studies have been performed on the eel.

In fish, water temperature is of primary importance because the body temperature of fish and the pathogen generation time are temperature-related. In the Japanese eel, agglutinating antibodies are readily produced at 16–19°C but not at 7–11°C.[21]

Use of bacterins

For immunization of eel against *E. tarda* infection, bacterins (formalin or heat killed cells or sonicated cells) were first tested. The methods of delivery used were immersion, injection, spray infusion and oral with food. Intramuscular (i.m.) injection has the disadvantage that it is difficult to inoculate a large number of fish. However, immersion and injection showed better results in protection, followed by spray infusion and least by oral administration (unpublished data).

The immersion method using formalin killed or heat killed cell preparations (Fig. 11.2) as antigens induced high antibody titre in 150 g eels at water temperatures between 20 and 30°C.[22] While the serum of immunized eels opsonized *E. tarda* cells, mortalities still occurred after challenge.

These results showed that, in the case of *E. tarda* infection, the protection is not strictly correlated with the agglutinating antibody titre. While results using bacterins were encouraging, a better level of protection was aimed for using semi-purified antigens. Therefore, investigations were carried out to identify the immunogen which produced protective immunity in the eel in the hope of improving the vaccines.

The search for protective antigens

The objective of the research was to investigate the usefulness of separated antigens or extracts from *E. tarda* (see Nature of Vaccines) and to determine their effectiveness using experimental challenge. As a general schedule for immunizing eels, the antigen preparations were injected on two occasions i.m. and the agglutinating antibody titre and/or passive haemagglutination antibody titre (using erythrocytes coated with the crude LPS antigen) were determined 3 weeks after the second immunization. Experimental challenges were performed by i.m. injection of viable live *E. tarda* strain EF-1 a few days after the determination of antibody titres in immunized eels and controls (Fig. 11.3).

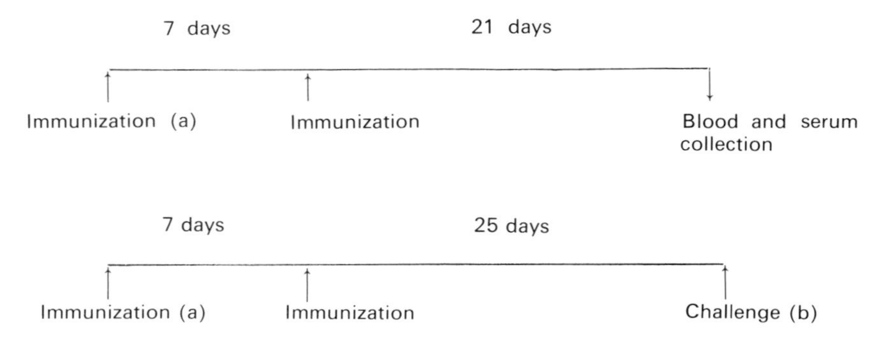

Fig. 11.3 Immunization and challenge schedule. (a) 90–95 g eels were injected i.m. (see Tables for doses), (b) eels were injected i.m. with *E. tarda* EF-1 viable cells. Eels were observed for 3 weeks after the challenge.

Preliminary studies[23] on the toxicity of endotoxin (lipopolysaccharide, LPS) and exotoxin of *E. tarda* EF-1 showed that the crude LPS obtained by the phenol–water method of Westphal and Jann[24] and the exotoxin from the culture filtrate were not highly toxic for the eel and therefore usable for immunization tests. In attempts to identify the protective antigens, formalin killed cells (FKC), exotoxin (culture filtrate) and crude LPS preparations (Fig. 11.4) obtained from *E. tarda* strain EF-1 were injected i.m. into 90–95 g eels in aquaria at constant temperature of 25°C.[25] All the preparations showed increase of antibody titres but only the group immunized with the crude LPS showed a high survival to the experimental challenge, while FKC and exotoxin preparations failed to induce protection (Table 11.1). Although LPS is contained in the FKC preparation, it is believed that the crude LPS preparation contains more LPS than the other preparations.

Table 11.1 *Immune response of eel against* Edwardsiella tarda *antigens*

Immunogen	Antigen dose	Antibody titre (1:)*	Challenge dose	Mortality %
Exotoxin	1.5 mg/eel**	6591		100
FKC	60.0 mg/eel	5904		100
LPS	1.0 mg/eel	3899	1.5×10^8 cells/eel (i.m.)	20
Contol		15		100

* Geometric mean of agglutinating antibody titre against formalin killed cells, determined before challenge.
** 90–95 g eels (*Anguilla japonica*), maintained at constant water temperature of 25°C.

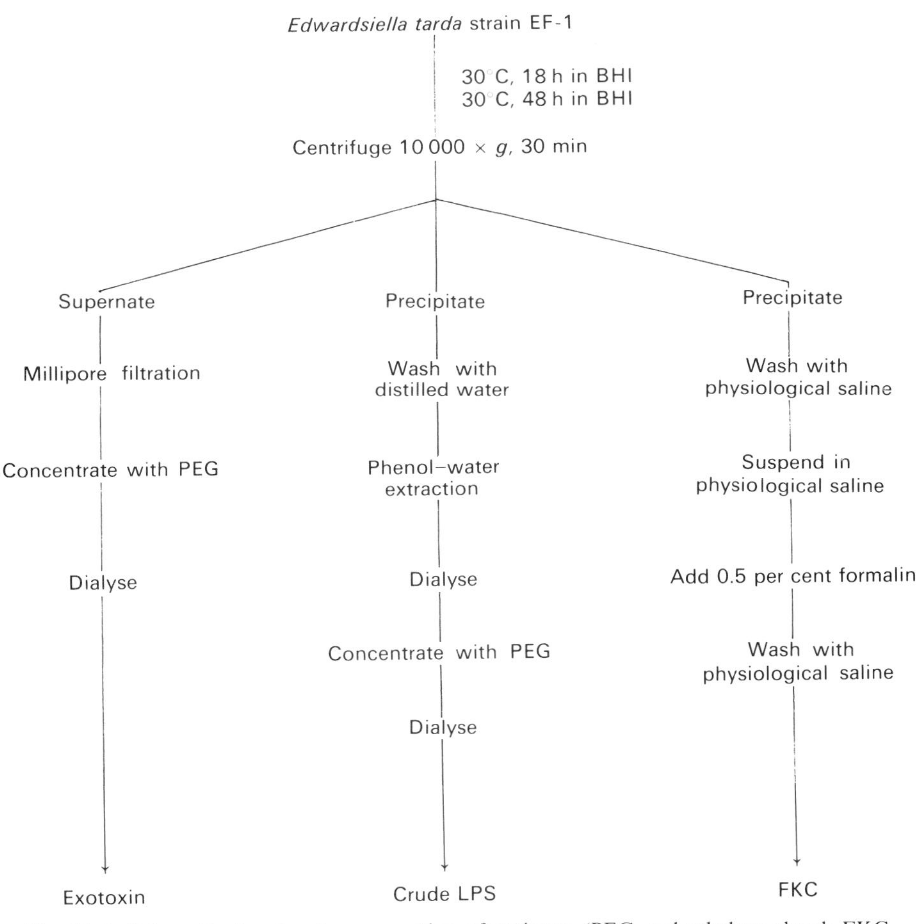

Fig. 11.4 Schematic drawing for preparation of antigens. (PEG, polyethylene glycol; FKC formalin killed cells).

In order to clarify the protective properties of *E. tarda* LPS, crude LPS, purified LPS, crude polysaccharide (PS) and lipid A prepared from strain EF-1 (Fig. 11.5) were injected i.m. into 90–95 g eels maintained in aquaria at 25°C.[26] The eels immunized with the four preparations showed an increase in agglutinating antibody titre in comparison to the controls (Table 11.2). The highest titre was recorded in eels immunized with the crude LPS preparation and the lowest with the lipid A. However, the highest survival to a strong experimental challenge was in the group immunized with the crude PS preparation and some resistance was demonstrated in the eels immunized with crude LPS. The eels immunized with purified LPS and lipid A preparations did not show protection to the challenge.

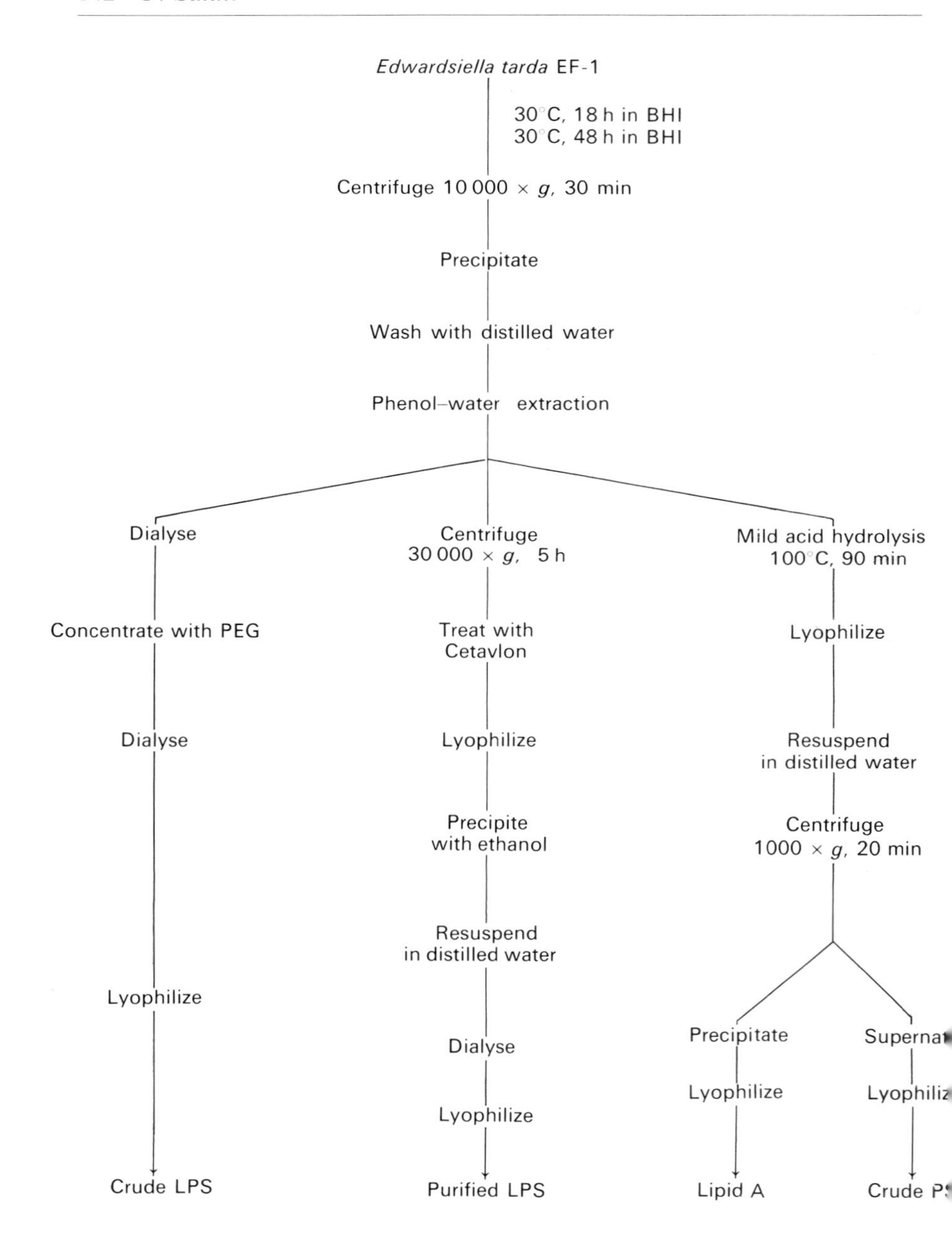

Fig. 11.5 Schematic drawing for preparation of antigens.

Table 11.2 *Immune response of eel against* Edwardsiella tarda *lipopolysaccharide*

Immunogen	Antigen dose	Antibody titre (1:)*	Challenge dose	Mortality %
Crude LPS	1.0 mg/eel**	5043		100
Purified LPS	0.25 mg/eel	2048		100
Polysaccharide	0.18 mg/eel	1098	1.9×10^8 cells/eel (i.m.)	0
Lipid A	0.07 mg/eel	275		100
Control		5		100

*Geometric mean of agglutinating antibody titre against formalin killed cells, determined before challenge.
**90–95 g eels (*Anguilla japonica*), maintained at constant water temperature of 25°C.

Chemical analysis showed a high degree of heterogeneity in the crude LPS with a small amount of protein and high content of nucleic acid. After further purification, the nucleic acids were eliminated and the proteins were reduced in the purified LPS preparation, but not completely eliminated. As a result, it was concluded that the main protective substance in the *E. tarda* LPS was the polysaccharide component. However, the immunogenicity of LPS or PS preparations may have been increased by the presence of small quantities of protein and nucleic acid remaining in the preparations since crude preparations of LPS induced better protection than highly purified preparations.

Table 11.3 *Chemical analysis of* Edwardsiella tarda *EF-1 purified lipopolysaccharide*

Chemical composition		
	Protein	3.6%
	Sugar	22.0%
	Fatty acid	8.3%
Monosaccharide composition		
	KDO	4.1%
	Pentose	1.1%
	Hexose	12.0%
	Galactose	0.5%
	Glucose	0.6%
	6-deoxyhexose	0.6%
	Heptose	3.9%
	Amino sugar	13.3%
Principal constituents of fatty acids		
	β-hydroxymyristic acid	
	Myristic acid	
	Palmitic acid	

As the LPS of *E. tarda* was shown to produce immunity in eel against experimental infection, a chemical analysis of the LPS was performed.[27] The results of the gas–liquid chromatography analysis of fatty acids showed that the principal constituents were β-hydroxymyristic acid, myristic acid and palmitic acid (Table 11.3). The composition of fatty acids, amino sugar, glucose, galactose, heptose and 2-keto-3-deoxyhexose (KDO) showed a taxonomic relation between *E. tarda* and *Salmonella*. The presence of non-hydroxylated fatty acids was a possible cause for concern since it has been demonstrated, using an anticomplement activity test, that preparations containing these were immunosuppressive.[28]

Further work was therefore conducted to study the importance of polysaccharide as a determinant of immune protection in the eels particularly after the elimination of lipid A from the immunogen.[29] *E. tarda* crude LPS, purified PS and whole cell without lipid (CWL) preparations (Fig. 11.6) were injected i.m. into 90–95 g eels at constant water temperature of 25°C. The highest survival rate after experimental challenge was in eels immunized with crude LPS preparation (Table 11.4) and good resistance was shown in eels immunized with the whole cell without lipid preparation. The eels immunized with purified PS did not show significant protection. Chemical analysis showed the PS to be highly purified and perhaps this was the reason for the loss of immunogenicity and effectiveness of the preparation. The results also suggested that the elimination of the lipid component from the whole cell may be an easy way to obtain an effective vaccine preparation against *E. tarda* infection. They certainly indicated that the lipid does not stimulate protective immunity in eels, but it was not clear whether the lipid suppressed the protective immune response or not.

Table 11.4 *Immune response of eel against* Edwardsiella tarda *vaccine preparations*

Immunogen	Antigen dose	Antibody titre (1:)*	Challenge dose	Mortality %
Crude LPS	1.0 mg/eel**	676		12.5
CWL	10.0 mg/eel	4705	1.1×10^7 cells/eel	31.6
Purified PS	0.04 mg/eel	388	(i.m.)	52.6
Control		4		70.0

*Geometric mean of agglutinating antibody titre against formalin killed cells, determined before challenge.
**90–95 g (*Anguilla japonica*), maintained at constant water temperature of 25°C.

Studies[30] were therefore performed to investigate the possible immunosuppressive properties of the lipid from *E. tarda*. Crude LPS, lipid extracted from the whole cell, FKC and CWL preparations (Fig. 11.7) were injected i.m. in 90–95 g eels maintained at constant water temperature of 25°C. In this experiment, the eels immunized with CWL preparation showed lower

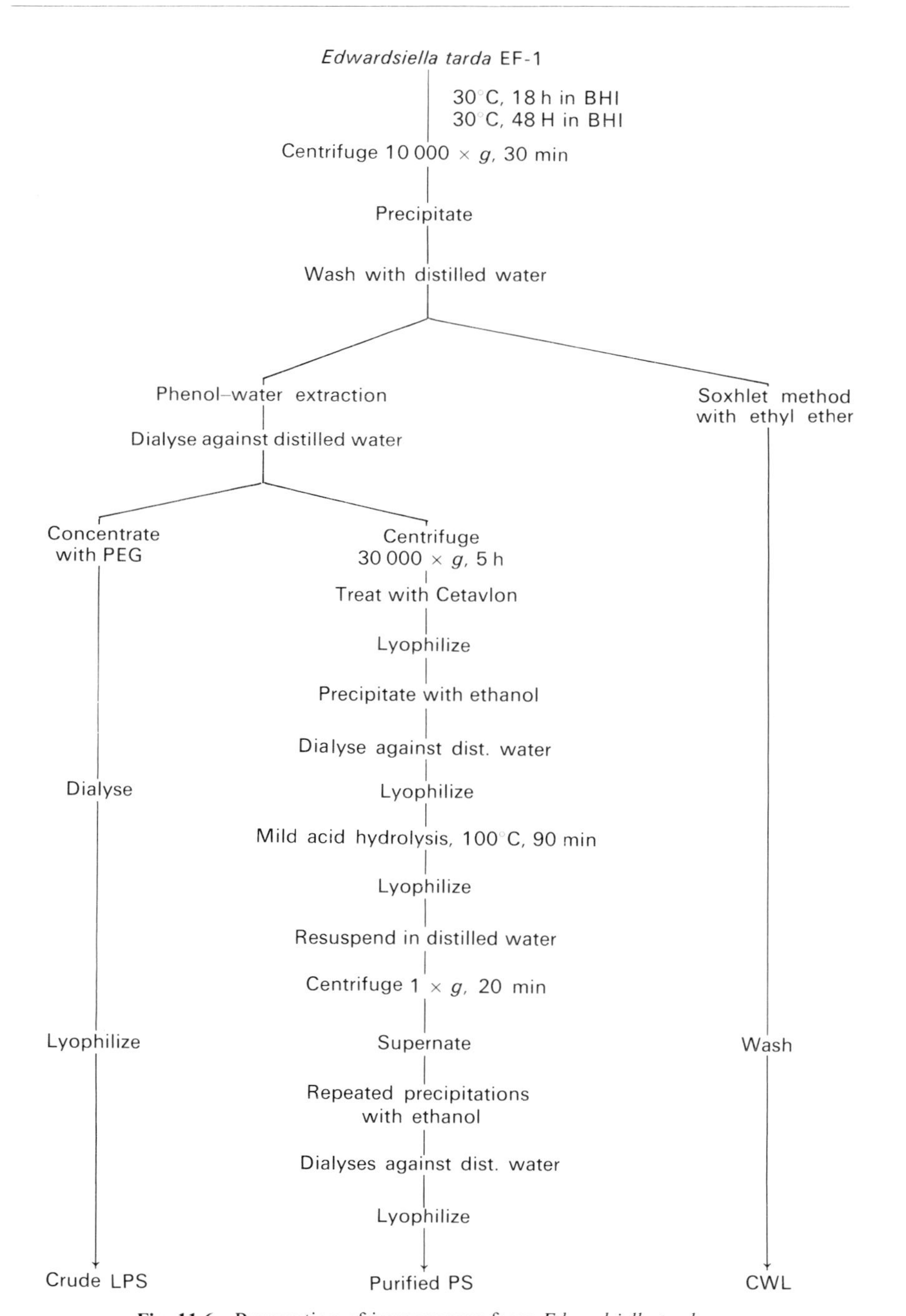

Fig. 11.6 Preparation of immunogens from *Edwardsiella tarda*.

protection and antibody titres than those obtained in a previous study.[29] This may have been due to possible heterogeneity of the eels used and/or to differences in immunogenicity of the preparations because the extraction of the lipid was carried out using different methods (see Figs. 11.6 and 11.7). The eels immunized with the lipid preparation did not show any protection against challenge and the highest survival rate was again in eels immunized with the crude LPS preparation (Table 11.5).

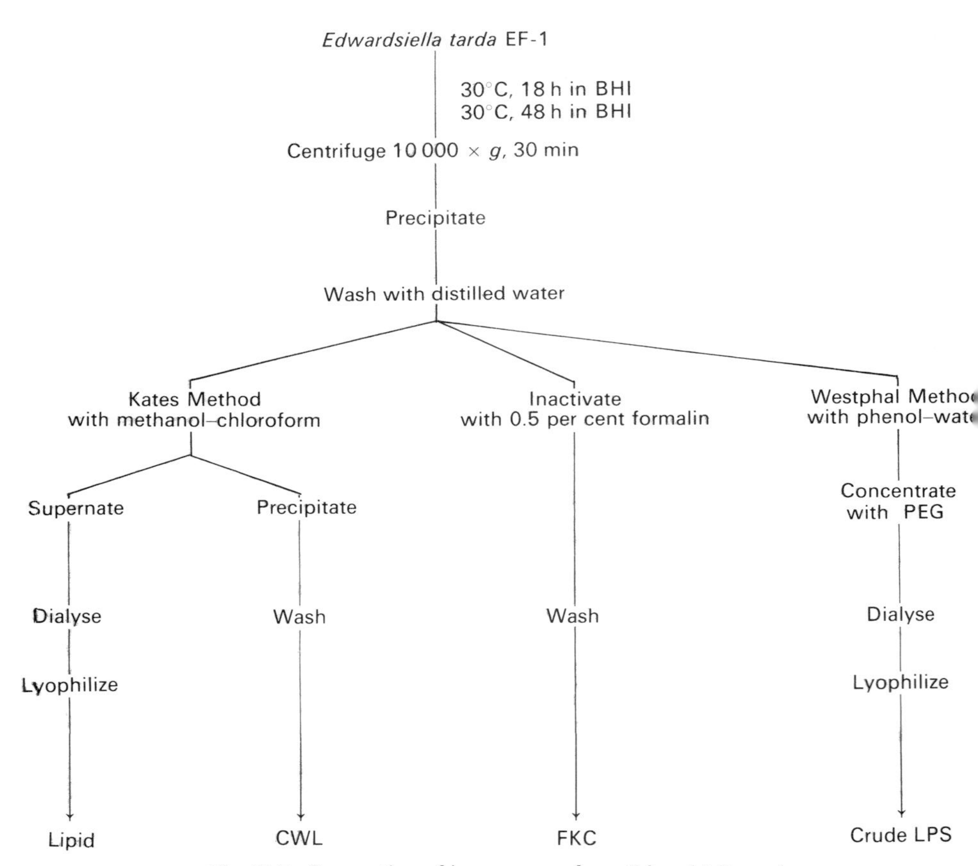

Fig. 11.7 Preparation of immunogens from *Edwardsiella tarda*.

The antibody titre of the eels immunized with lipid was very low and also the *in vitro* phagocytic activity of the total blood was significantly lower than that found in the controls (Table 11.6). This result strongly suggests that the lipid of *E. tarda* does act as an immunosuppressant in the eel at least as far as the phagocytic defences are concerned. The eels immunized with crude LPS

Table 11.5 *Immune response of eel to* Edwardsiella tarda *lipid*

Immunogen	Antigen dose	Antibody titre (1:)*	Challenge dose	Mortality %
Lipid	1.0 mg/eel**	13		100
LPS	1.0 mg/eel	1218		85
KFC	10.0 mg/eel	1024	2.4×10^7 cells/eel	100
CWL	10.0 mg/eel	431	(i.m.)	95
Control		7		100

*Geometric mean of agglutinating antibody titre against formalin killed cells, determined before challenge.
**90–95 g eels (*Anguilla japonica*), maintained at constant water temperature of 25°C.

again showed high antibody titre and significantly enhanced *in vitro* phagocytosis activity (Table 11.6). Further studies on total blood phagocytosis showed that the *in vitro* phagocytosis activity of eels immunized with crude LPS was significantly higher than controls from the third hour of incubation.[31] These results confirm the capacity of the crude LPS to enhance phagocytic activity of eel leucocytes and to increase serum antibody titres, the combined activity of which may protect the eel from natural infection by *E. tarda*.

Table 11.6 In vitro *phagocytosis of total blood from eels immunized with* Edwardsiella tarda *crude lipopolysaccharide (LPS) and lipid preparations*

Immunization group	Phagocytosis* activity	Phagocytic** index
LPS	74.5 ± 3.8	8.7 ± 2.1
Lipid	10.5 ± 4.9	2.0 ± 0.7
Control	43.5 ± 6.5	8.7 ± 1.9

*Phagocytosis activity: per cent of cells showing phagocytic activity.
**Phagocytic index: number of bacteria ingested/phagocyte in a 24 h incubation period.

Recently,[32] studies in the immune response of red sea bream injected i.m. once with *E. tarda* FKC and crude LPS preparations (Fig. 11.4) showed increased serum agglutinating antibody titres from the first week after immunization, particularly in the fish immunized with FKC preparation. Assays of the *in vitro* phagocytic index of the total blood showed enhanced phagocytosis from the first week after immunization in both crude LPS and FKC immunized red sea breams. The authors suggest that opsonization by antibody and phagocytosis may be one important mechanism of defence against *E. tarda* infection and if these two factors can be affected independently, it may explain why protection is not always strictly correlated to the agglutinating antibody titre.

LIMITATIONS OF PRESENT VACCINES

As stated above, in order to obtain an immune response in eel, it is necessary that the water temperature is higher than 16°C. Fortunately, this factor raises no problem since *E. tarda* has low pathogenicity in eels at temperatures below 24°C. While immersion and injection vaccination of anguillettes has been shown to provide protection, a major limitation of the present knowledge is the effectiveness in elvers which is the stage principally at risk. It was demonstrated that glass eel immunized by immersion method with FKC preparation did not show any protection to experimental infection with live *E. tarda*. Obviously, more work is required here. While injection immunization of anguillettes can achieve protection, more information is required on immersion vaccination with antigen extracts before commercial vaccination at this stage of the life cycle can be contemplated. Long durability of protection may not be important as adult eels are naturally more resistant to the infection than anguillettes.

LIMITATIONS REGARDING ANTIGEN PREPARATION

It is possible to suggest that the crude polysaccharide achieves the best immune protection to the infection. The crude LPS, with its characteristics of heterogeneity, increases antibody titres and enhances total blood phagocytosis activity. Even if less protective than crude polysaccharide, crude LPS is a good immunogen, is easier to prepare and may be more suitable for preparation of vaccines. However, the possibility of induced protection by immersion method using a crude LPS preparation has not been fully tested. This is essential as a commercial vaccine must be delivered in this manner. More work is required on the duration and specificity of the protection, though it may be presumed that as the antibodies produced on immunization can effectively opsonize *E. tarda* cells in *in vitro* phagocytic test, these antibodies are specific. This is important since LPS is known to non-specifically stimulate many defence parameters but protection which results from this is likely to be short lived and of course has no memory component.

The study and the development of new vaccines using adjuvants or synthetic substances should be performed. However, the cost of the research, the limited period of use in aquaculture and above all, the cost of obtaining product licences from government authorities depress the development of new vaccines. For example in Japan, even if research is well developed in the universities, the principal obstacle is at legislative level where it is difficult to obtain permission from the Ministry of Health for practical use in aquaculture.

CONCLUSIONS

While vaccination studies to date have been experimental, on animals a little older than those at major risk and using the injection route, there are grounds for optimism concerning the development of a commercially useful vaccine.

Bacterins which are simple and cheap to produce on a large scale do provide protection to anguillettes and while the protection (by experimental challenge) is not complete, it may prove useful enough to farms with a severe *Edwardsiella* problem. Field trials have not been conducted but they may be justified on a small scale.

Certain antigen extracts, especially the crude polysaccharide and lipopolysaccharide, appear to be better immunogens but large scale production may be non-economic for a small market. There is still much to be done concerning the feasibility of immunizing elvers by immersion. Now that quite good vaccines have been developed when tested experimentally in anguillettes, future work should be directed to their usefulness in elvers. Edwardsiellosis is a disease of warmwater fish culture and is mainly a problem for eel farms in Japan and Taiwan. However, the disease has a wide range of hosts and as the culture of warmwater species expands in the tropics and subtropics so the problem from edwardsiellosis may also expand. If this proves to be the case there will be a greater commercial interest in producing a vaccine but at the moment this incentive is lacking.

REFERENCES

1 Hoshina, T. (1962). On a new bacterium, *Paracolobactrum anguillimortiferum* n. sp. *Bull. Japan. Soc. Sci. Fish.*, **28** (2), 162–164.
2 Wakabayashi, H. and Egusa, S. (1973). *Edwardsiella tarda* (*Paracolobactrum anguillimortiferum*) associated with pond-cultured eel disease. *Bull. Japan. Soc. Sci. Fis.*, **39** (9), 931–936.
3 Amandi, A., Hiu, S. F., Rohovec, J. S. and Fryer, J. L. (1982). Isolation and characterization of *Edwardsiella tarda* from fall chinook salmon (*Oncorhynchus tshawytscha*). *Appl. Environ. Microbiol.*, **43**, 1380–1384.
4 Bockemuhl, J., Pan-Urai, R. and Burkhardt, F. (1971). *Edwardsiella tarda* associated with human disease. *Path. Microbiol.*, **37**, 393–401.

5 Kusuda, R., Toyoshima, T., Iwamura, Y. and Sako, H. (1976). *Edwardsiella tarda* from an epizootic of mullets (*Mugil cephalus*) in Okitsu Bay. *Bull. Japan. Soc. Sci. Fish.*, **42**, (3) 271–275.
6 Kusuda, R., Itami, T., Munekiyo, M. and Nakajima, H. (1977). Characteristics of *Edwardsiella* sp. from an epizootic of cultured crimson sea breams. *Bull. Japan. Soc. Sci. Fish.*, **43** (2), 129–134.
7 Yasunga, N., Ogawa, S. and Hatai, K. (1982). Characteristics of the fish pathogen *Edwardsiella* isolated from several species of cultured marine fishes. *Bul. Nagasaki Pref. Inst. Fish.*, **8**, 57–65.
8 Nakatsugawa, T. (1983). *Edwardsiella tarda* isolated from cultured young flounder. *Fish Pathol.*, **18** (2), 99–101.
9 Sakazaki, R. (1967). Studies on the Asakusa group of *Enterobacteriaceae* (*Edwardsiella tarda*). *Japan. J. Med. Sci. Bio.*, **20**, 205–212.
10 White, F. H., Simpson, C. F. and Williams, L. E. Jr (1973). Isolation of *Edwardsiella tarda* from aquatic animal species and surface waters in Florida. *J. Wildlife Dis.*, **9**, 204–208.
11 Van Damme, L. R. and Vandepitte, J. (1980). Frequent isolation of *Edwardsiella tarda* and *Plesiomonas shigelloides* from healthy Zairese freshwater fish: a possible source of sporadic diarrhoea in the tropics. *Appl. and Env. Microbiol.*, **39** (3), 475–479.
12 Ishihara, S. and Kusuda, R. (1981). Experimental infection of elvers and anguillettes with *Edwardsiella tarda* bacteria. *Bull. Japan. Soc. Sci. Fish.*, **47** (8), 999–1002.
13 Wakabayashi, H., Kanai, K. and Egusa, S. (1976). Ecological studies of fish pathogenic bacteria in eel farm-I. Isolation of aerobic bacteria from pond water. *Fish Pathol.*, **11** (2), 63–66.
14 Egusa S. (1978). *Gyo no kansensho* (*Infectious Diseases in Fish*). Tokyo: Koseisha Koseikaku, pp. 164–168.
15 Miyazaki, T. and Egusa, S. (1976). Histopathological studies of edwardsiellosis of the Japanese eel (*Anguilla japonica*). Suppurative interstitial nephritis form. *Fish Pathol.*, **11**, 33–43.
16 Miyazaki, T. and Egusa, S. (1976). Histopathological studies of edwardsiellosis of the Japanese eel (*Anguilla japonica*). Suppurative hepatitis form. *Fish Pathol.*, **11**, 67–75.
17 Ullah, Md-A. and Arai, T. (1983). Pathological activities of the naturally occurring strains of *Edwardsiella tarda*. *Fish Pathol.*, **18** (2), 65–70.
18 Ullah, Md-A. and Arai, T. (1983). Exotoxic substances produced by *Edwardsiella tarda*. *Fish Pathol.*, **18** (2), 71–75.
19 Kusuda, R. and Ishihara, S. (1981). The fate of *Edwardsiella tarda* bacteria after intramuscular injection of eels. *Anguilla japonica*. *Bull. Japan. Soc. Sci. Fish.*, **47** (4), 475–479.
20 Edwards, P. R. and Ewing, W. H. (1972). *Identification of Enterobacteriaceae*. Third edn. Minneapolis: Burgess Publish. Co.
21 Muroga, K. and Egusa, S. (1969). Immune response in Japanese eel to *Vibrio anguillarum*. I. Effects of temperature on agglutinating antibody production in starved eels. *Bull. Japan. Soc. Sci. Fish.*, **35**, 868–874.
22 Song, Y. L. and Kou, G. H. (1981). The immunoresponses of eels (*Anguilla japonica*) against *Edwardsiella anguillimortifera* as studied by the immersion method. *Fish Pathol.*, **15** (3), 249–255.

23 Salati, F. (1985). Immunogenicity of *Edwardsiella tarda* antigens in the eel (*Anguilla japonica*). *Riv. It. Piscic. Ittiop.*, **20**(1), 12–26.
24 Westphal, O. and Jann, K. (1965). Bacterial lipopolysaccharides. Extraction with phenol–water and further applications of the procedure. In: *Methods in Carbohydrate Chemistry*, ed. R. L. Whistler, J. N. BeMiller and M. L. Wolfrom, Vol. V, New York: Academic Press, pp. 83–91.
25 Salati, F., Kawai, K. and Kusuda, R. (1983). Immuno-response of eel against *Edwardsiella tarda* antigens. *Fish Pathol.*, **18** (3), 135–141.
26 Salati, F., Kawai, K. and Kusuda, R. (1984). Immune response of eel to *Edwardsiella tarda* lipopolysaccharide. *Fish Pathol.*, **19** (3), 187–192.
27 Salati, F. and Kusuda, R. (1985). Chemical composition of the lipopolysaccharide from *Edwardsiella tarda*. *Fish Pathol.*, **20** (2/3), 187–191.
28 Tanamoto, K., Zähringer, U., McKenzie, G. R., Galanos, C., Rietschel, E. T., Lüderitz, O., Kusumoto, S. and Shiba, T.(1984). Biological activities of synthetic lipid A analogs: pyrogenicity, lethal toxicity, anticomplement activity and induction of gelation of *Limulus* amoebocyte lysate. *Infect. Immun.*, **44** (2), 421–426.
29 Salati, F. and Kusuda, R. (1985). Vaccine preparations used for immunization of eel *Anguilla japonica* against *Edwardsiella tarda* infection. *Bull. Japan. Soc. Sci. Fish.*, **51** (8), 1233–1237.
30 Salati, F. and Kusuda, R. (1986). Immune response of eel to *Edwardsiella tarda* lipid. *Fish Pathol.*, **21** (3), 201–205.
31 Salati, F., Ikeda, Y. and Kusuda, R. (1987). Effect of *Edwardsiella tarda* lipopolysaccharide immunization on phagocytosis in the eel. *Bull. Japan. Soc. Sci. Fish.*, **53** (2), 201–204.
32 Salati, F., Hamaguchi, M. and Kusuda, R. (1987). Immune response of red sea bream to *Edwardsiella tarda* antigens. *Fish Pathol.*, **22** (2), 93–98.

12

Vaccination against *Edwardsiella ictaluri*

J. A. Plumb

THE DISEASE PROBLEM

The bacterium *Edwardsiella ictaluri* is the aetiological agent of enteric septicaemia of channel catfish (ESC) which has become the most serious infectious disease problem of cultured channel catfish (*Ictalurus punctatus*) since its original discovery in 1978.[1,2] In 1985, 1042 cases of *E. ictaluri* were reported from the southeastern United States and constituted 28 per cent of the diseases of all fish species reported during the year (A. J. Mitchell, Stuttgart, Ark, pers. commun.). Severity of the cases ranged from low ($\leqslant$ 5%) to severe ($\geqslant$50%) mortality and cost the catfish industry millions of dollars in fish and medicated feed.

Channel catfish are grown primarily in the warmer southern latitudes of the United States, but culture of this species is expanding into Europe, Central America and Asia. The potential for ESC to occur in other parts of the world is significant.

The channel catfish is the principal species affected by *E. ictaluri*, but it has also been isolated from two aquarium species, the danio (*Danio devario*) and the green knife fish (*Eigemannia vireens*), and the walking catfish (*Clarias batrachus*) in Thailand. It has been shown experimentally that other species of fish are not as susceptible to *E. ictaluri* as the channel catfish. Low susceptibility to *E. ictaluri* occurs in the blue tilapia (*Sarotherodon aureus*), golden shiner (*Notemigonus crysoleucas*), largemouth bass (*Micropterous salmoides*), bighead carp (*Aristichthys nobilis*) and the European catfish (*Silurus glanis*). Whether or not the non-ictalurid species that are susceptible to *E. ictaluri* can serve as carriers and infect more susceptible populations is not known.

Experimental work on age susceptibility of channel catfish to *E. ictaluri* has not been done, but cases have been reported involving fingerling size to adult fish. However, production-size fish (over 100 g) appear to be more severely affected. ESC is generally a seasonal disease occurring when water temperatures are between 18°C and 28°C (Fig. 12.1), but the disease is not

FISH VACCINATION
ISBN 0-12-237485-1

strictly confined to this temperature range. Thus, most cases occur in May–June and September–November, with noticeable reductions in the winter and summer. Hawke[1] determined that the optimum *in vitro* growth temperature for *E. ictaluri* is 25°C to 30°C, the normal temperature range in catfish ponds in the south-eastern United States during the spring and fall.

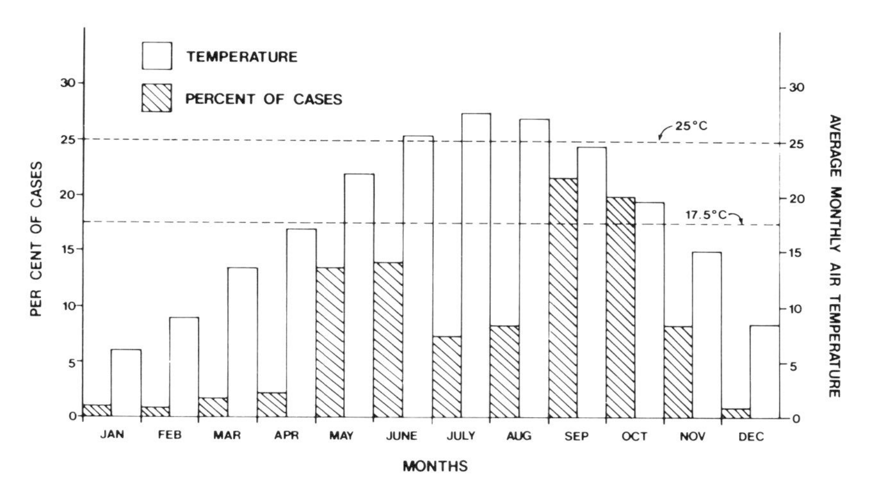

Fig. 12.1 Seasonal occurrence of 602 *Edwardsiella ictaluri* cases in cultured channel catfish in Alabama and Mississippi. Part of these data were supplied by T. E. Schwedler and the Mississippi Cooperative Extension Service. The bacteria survive for longer periods in mud between 18 and 25°C than lower or higher temperatures (see Fig. 12.2).

Empirical epizootiological information concerning the relationship of *E. ictaluri* infections to stocking density, water quality, holding facilities, etc. is not available. However, as is the case with most infectious diseases of channel catfish, ESC appears to be more severe in high density populations and when water quality, such as oxygen concentrations and organic load, is less than ideal. ESC has been diagnosed in channel catfish populations in ponds, raceways and tanks, but seldom in wild populations.

Transmission of *E. ictaluri* from fish to fish occurs through the water. The disease can be contracted by water-borne exposure, injection or by oral administration, however, Miyazaki and Plumb[3] suggested that the nares are the primary site of invasion of the fish.

Little is known about the carrier state of *E. ictaluri* in catfish, but the bacterium has been isolated from the brain of asymptomatic catfish.

Originally it was believed that *E. ictaluri* was an obligate pathogen because it survived for only 8 days in pond water.[1] However, Plumb and

Quinlan[4] showed that *E. ictaluri* survived for over 95 days in pond bottom mud when incubated at 25°C and for at least 45 days at 18°C (Fig. 12.2), therefore bottom mud could be a reservoir for *E. ictaluri*.

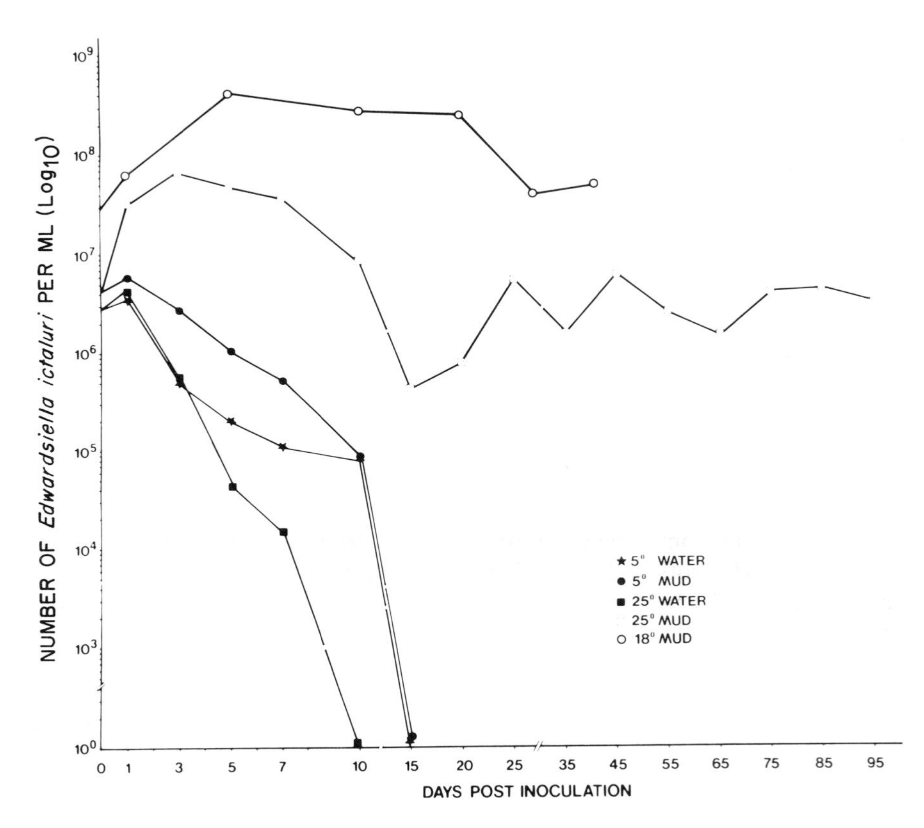

Fig. 12.2 Survival of *Edwardsiella ictaluri* in pond water and mud when incubated at 5°C, 18°C and 25°C. (From Plumb and Quinlan[4]).

All US isolates so far examined are identical and there is strong serological homogeneity between several US *E. ictaluri* isolates.[5] There is very little biochemical heterogeneity between US *E. ictaluri* and Thailand isolates, but agglutination tests have indicated there may be some serological difference. Generally, the *E. ictaluri* is a homogeneous group and only one strain is presently recognized.

THE IDEAL VACCINE AND ITS STRATEGIC USE

The vaccine should be applicable to routine hatchery practices and would best be used as an immersion for fingerling channel catfish from early spring to fall when they are being stocked into grower ponds. Efficacy of the vaccine should reduce the mortality to the point that chemotherapy is not required. Normal production cycle of catfish in grower ponds is approximately 1 year. Therefore, protection should preferably cover this whole period. Where broodstock must be immunized, an injectable vaccine would be acceptable.

Due to the seasonal nature of ESC (spring and fall), protection should occur during these periods. Although grower ponds may be stocked throughout the year, most ponds are stocked in the fall and winter. If the catfish build immunity during the winter, it would be appropriate to vaccinate fingerlings in the fall providing protection for the following spring and hopefully persisting into the fall.

NATURE OF THE VACCINES

Only experimental *E. ictaluri* vaccines are in existence now and none are available from commercial sources. *E. ictaluri* is grown in Brain Heart Infusion (BHI) broth at 28°C with either agitation or sterile aeration for 24 h at which time the bacterial concentration is approximately 1×10^9 cells per ml. Although not reported in the literature, commercial fermenters should be very useful in growing *E. ictaluri* for vaccine preparation. Most of the experimental vaccination work has been done with bacterins composed of formalin killed whole cells or sonicated cultures following formalin treatment.[6] Formalin is added at 1 per cent of the culture in the BHI broth and allowed to stand overnight and then sonicated at 20 s bursts for 1 min. Another preparation included centrifuging the cells and disrupting them in a French press at 18 000 psi for 2 min. Also, a toxoid vaccine with a lipopolysaccharide (LPS) extract was used by Saeed and Plumb.[7]

Vaccines are applied to channel catfish as either injection or an immersion bath. Injections are 0.1 ml of vaccine without dilution intraperitoneally regardless of fish size. For immersion, the vaccines are diluted in water at 1:10 and fish immersed for 1 to 4 min. Optimum dilution of vaccine and time of immersion have not been determined.

The protective antigenic component of *E. ictaluri* is not known, although Saeed and Plumb[7] showed that the lipopolysaccharide (LPS) of the pathogen administered by injection is antigenic and, to a degree, protective.

VACCINE TRIALS

Lipopolysaccharide (LPS) vaccines

Limited experimental vaccinations for *E. ictaluri* have been carried out. Channel catfish that were vaccinated three times with *E. ictaluri* LPS in Freund's complete adjuvant (FCA) by injection were protected against challenge (3.3 per cent mortality) as opposed to vaccination with LPS alone (26.7 per cent mortality) or non-vaccinated controls (70 per cent mortality).[7] Fish vaccinated by single injection with LPS in FCA suffered 20 per cent mortality after challenge compared to 80 per cent mortality following single LPS injection without adjuvant or saline controls. Injection with FCA alone provided no protection. Vaccination by immersion in LPS was not strongly immunogenic or protective.

Bacterin vaccines

Plumb *et al.*[6] demonstrated that channel catfish vaccinated by injection with a formalin killed bacterin developed agglutinating antibodies when the fish were held for 30 days at 25°C followed by 30 days at 12°C. In fact, the antibody titres were higher in the 12°C fish than in fish held at 25°C for 60 days (Fig. 12.3). The fish held at 25°C and then 12°C showed stronger protective immunity to challenge at 25°C than fish held at 25°C for the entire study. In a subsequent study, channel catfish were either immersed in a sonicated or whole cell bacterin or injected with the same preparations and held at 25°C for 4 days prior to lowering the temperatures to 12°C, and 4 months after vaccination the fish were challenged at 25°C with virulent *E. ictaluri* by water-borne exposure.[6] Fish that were immersed in the sonicated vaccine suffered 12 per cent mortality and those injected with the sonicate had 42 per cent mortality, while the immersed and injected whole cell vaccinated fish had 25 per cent and 58 per cent mortality, respectively. Non-immunized control fish suffered 46 per cent mortality. These data indicate that a sonicated preparation of *E. ictaluri* is more protective than a whole cell preparation and that immersion in the vaccine is superior to injection.

In a more extensive field trial, 600 channel catfish fingerlings were vaccinated by immersion in a whole cell bacterin, a sonicated or a French press preparation of *E. ictaluri* in early November, held at 24°C for 3 days then stocked into wintering ponds (Plumb, unpublished data). At 30-day intervals, *E. ictaluri* agglutinating antibody titres were determined (Table 12.1). Over a 150-day period, unvaccinated fish had zero antibody titres but fish vaccinated with the sonicated, French press and whole cell preparations had developed antibody titres by 30 days and these persisted for the 150 days period (December to April). During this time, average monthly water temperatures ranged from 17°C in November to 7.2°C in January and up to 22.6°C in April. A suitable challenge to these vaccinated fish was not obtained; however, the data show that fall vaccinated channel catfish did develop an immune response during the winter.

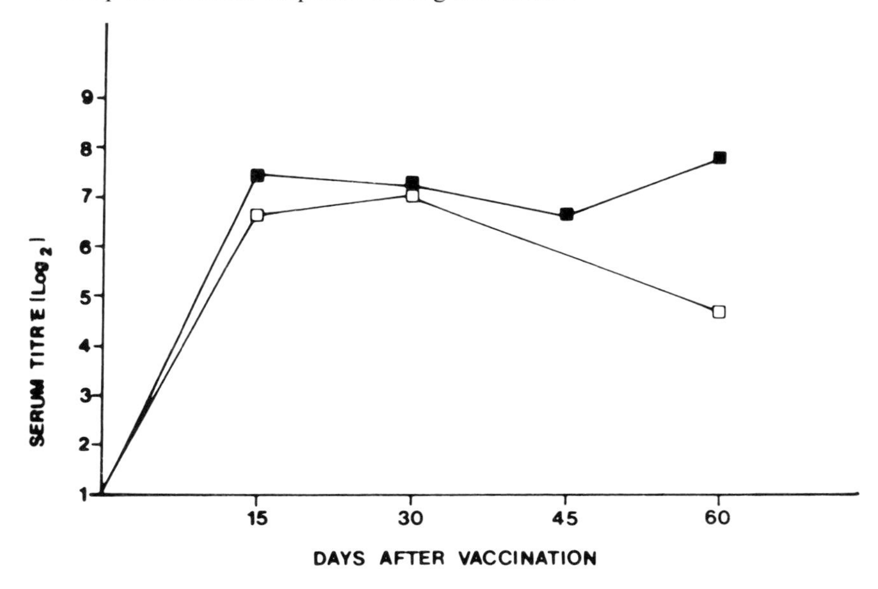

Fig. 12.3 Antibody titres of channel catfish vaccinated with *E. ictaluri* and held at 25°C for 30 days and then at 12°C for 30 days (■) or held at 25°C for 60 days (□). (From Plumb *et al.*[6]).

Table 12.1 *Agglutination antibody titres against* Edwardsiella ictaluri *of channel catfish following immersion vaccination and held in ponds over winter.*

	Mean Antibody titres[1] at Days Post Vaccination					
Vaccine Preparation	Day 0 1 Nov.	Day 30 1 Dec.	Day 60 2 Jan.	Day 90 1 Feb.	Day 120 3 Mar.	Day 150 2 Apr.
Unvaccinated	0	0	0	0	0	0
Sonicated	0	15.2 (4–32)	15.2 (4–32)	6.4 (4–8)	12 (8–16)	12 (4–16)
French press	0	4.8 (2–8)	4.8 (2–8)	7.2 (4–16)	12 (8–16)	3.6 (2–4)
Whole cell	0	9.6 (8.16)	5.2 (2–8)	4 (4–4)	14.4 (8–16)	5.6 (4–8)
Average water temperature	17.0	15.6	7.2	10.8	15.8	22.6

[1]Reciprocal of highest dilution where agglutination occurred.
Values in parentheses are the ranges of 20 fish sampled.

Table 12.2 *Relative per cent survival (RPS) of channel catfish immunized with* Edwardsiella ictaluri *by several methods and preparations of the antigen. RPS calculated by method described by Amend.*[8]

From Saeed and Plumb[7]			From Plumb *et al.*[6]		
Vaccination treatment	Per cent mortality[1]	RPS	Vaccination treatment	Per cent mortality[1]	RPS
Multiple injection			*Immersion*		
LPS + adjuvant	3.3	98.6	Sonicated	11.8	74.4
LPS only	36.7	47.6	Whole cell	24.6	46.6
Saline	70	–	*Injected*		
			Sonicated	46.7	9.4
Single injection			Whole cell	57.9	<0
LPS and adjuvant	20	75	Non-immunized	46	–
LPS only	80	0			
Saline	80	–			
Whole cell preparation					
Immersion	81.7	$\leqslant 0$			
Prehyperosmotic dip	78.3	$\leqslant 0$			
Injection and adjuvant	8.3	89.2			
Injection without adjuvant	76.7	–			

[1]Mortality determined following exposure to virulent *E. ictaluri*.

EFFECTIVENESS

Trials to vaccinate channel catfish against *E. ictaluri* have been successful enough to suggest that such vaccinations are possible. Table 12.2 summarizes the effectiveness of two experiments to vaccinate channel catfish.[6,7] Relative per cent survival (RPS)[8] was used to compare the efficacy of various methods of vaccination with several different vaccine preparations. Amend[8] suggested that an RPS of over 60 provides acceptable protection. Multiple and single injection of LPS with FCA resulted in RPS of 98.6 and 75.0, respectively, and was, therefore, a very effective immunogen but LPS injected alone was not (Table 12.2).[7] Injection of a whole cell preparation in FCA resulted in an RPS of 89.2. Direct immersion or prehyperosmotic dip with a whole cell preparation were not effectively protective.[7] Immersion of catfish in a sonicated preparation of *E. ictaluri* resulted in a protective RPS of 74.4 (Table 12.2), but immersion in the whole cell preparation or injection of either preparation was not protective.[6] These two studies indicate that injection of *E. ictaluri* is protective if an adjuvant is used. Data on the efficacy of immersion conflict and it is apparent that more research is required to resolve these questions. Longevity of immunity is not fully known. Although winter vaccinations by Plumb *et al.*[6] indicate at least a 4-month longevity, a longer protection is desirable. Sub-lethal challenges by Plumb *et al.*[6] showed that antibody levels of fish vaccinated 4 months previously accelerated to a higher level than unvaccinated fish, indicating that a natural exposure to *E. ictaluri* may serve as a booster.

Antibody from catfish against *E. ictaluri* are specific for this species and show no cross-reactivity with *E. tarda*, but challenge of *E. ictaluri* immunized fish with *E. tarda* has not been done.

LIMITATIONS OF PRESENT VACCINES

Present vaccines are limited to use on fingerling channel catfish in experimental conditions, and since they consist of killed bacteria they pose no health threat to fish populations. The earliest age for vaccination is not known; however, due to potential numbers of fish that could possibly be vaccinated, fish less than 10 cm in length would be the most desirable size to vaccinate by immersion. Immersion for 2 min in a vaccine is immunogenic, but a shorter exposure time is desirable and this must be established. A 1:10 dilution of a bacterin containing approximately 10^9 cells per ml is immunogenic, but a greater dilution would be advantageous thus reducing the cost of vaccination. The optimum dilution (cost and effectiveness) of the bacterin must also be determined. Complete protection against *E. ictaluri* has not been demonstrated, but on several occasions no more than 10 per cent of experimentally challenged vaccinated fish have died. The need for booster vaccinations has been shown by several trials, especially those of Saeed and Plumb[7] when using *E. ictaluri* LPS as an injectable vaccine. It is probable that natural exposure to *E. ictaluri* may serve as an effective booster and in the field situation it may not be necessary to remove fish from a pond simply for purposes of applying a booster vaccination.

OVERCOMING LIMITATIONS

The present limitations of *E. ictaluri* vaccines are the equivocal protectiveness, longevity of immunity and elimination of carriers. It has not been conclusively demonstrated that vaccination with crude vaccine or LPS of *E. ictaluri* fully protect the fish. Length of immunity and the need for booster vaccinations have not been resolved. In conjunction with this problem, it is not known if vaccination will eliminate carrier fish or if vaccinated fish can become carriers when exposed to the pathogen.

It appears that vaccination by immersion is a feasible method of immunizing channel catfish against *E. ictaluri*. This method is ideal for vaccinating fingerling fish, but the optimum size fish to immunize has not been determined. Routine vaccination of fish greater than 100 g (production size) would seem impractical and cost prohibitive.

COMMERCIAL PROSPECT

A vaccine for *E. ictaluri* has excellent commercial potential in view of the fact that ESC is a very serious and wide-spread problem in the expanding and already extensive catfish industry. To be practical, the commercial vaccine must be (a) cost effective, (b) convenient in its application procedure and (c) provide long term protection.

Initial development and production costs will be moderate, because some preliminary research has been done. Additional experimentation would likely be done by the vaccine manufacturer. Legal and licensing restrictions should not pose any unusual problems because the vaccine will be a killed preparation. While some aquarium fish species have been infected with *E. ictaluri*, the disease has not been shown to be a serious problem. With the detection of *E. ictaluri* in walking catfish in Thailand and the establishment of channel catfish in Europe and other parts of the world, the possibility of an international market for an *E. ictaluri* vaccine enhances the potential of commercial production.

CONCLUSIONS

Of the many diseases of channel catfish, ESC is the most likely to be controlled by prophylactic vaccination. *E. ictaluri* is an excellent antigen, it is easy to grow in a large volume, and basically ESC is confined to one species of fish. Although there is much work to be done in developing a vaccine for *E. ictaluri*, we know that it is immunogenic in an immersion application. The need is there, and the demand is there; however, a commercial vaccine producing company must assume the leadership in its development.

REFERENCES

1 Hawke, J. P. (1979). A bacterium associated with disease of pond cultured channel catfish, *Ictalurus punctatus*. *J. Fish Res. Board Can.*, **36**, 1508–1512.

2 Hawke, J. P., McWhorter, A. C., Steigerwalt, A. G. and Brenner, D. J. (1981). *Edwardsiella ictaluri* sp. nov., the causative agent of enteric septicemia of catfish. *Int. J. Syst. Bacteriol.,* **31**, 396–400.
3 Miyazaki, T. and Plumb, J. A. (1985). Histopathology of *Edwardsiella ictaluri* in channel catfish, *Ictalurus punctatus* (Rafinesque). *J. Fish Dis.,* **8**, 389–392.
4 Plumb, J. A. and Quinlan, E. E. (1986). Survival of *Edwardsiella ictaluri* in pond water and bottom mud. Progr. Fish-Cult., **48**, 212–214.
5 Rogers, W. A. (1981). Serological detection of two species of *Edwardsiella* infecting catfish. Intern. Symp. Fish Biol: Serodiog. and vaccines. *Develop. Biol. Stand.,* **49**, 169–172.
6 Plumb, J. A., Wise, M. L. and Rogers W. A. (1986). Modulary effects of temperature on antibody response and specific resistance to challenge of channel catfish, *Ictalurus punctatus,* immunized against *Edwardsiella ictaluri. Vet. Immunol. Immunopathol.,* **12**, 297–304.
7 Saeed, M. O. and Plumb, J. A. (1986). Immune response of channel catfish to the lipopolysaccharide and whole cell *Edwardsiella ictaluri* vaccines *Dis. Aquat. Org.,* **2**, 21–25.
8 Amend, D. F. (1981). Potency testing of fish vaccines. Intern. Symp. Fish Biol.: Serodiog. and Vaccines. *Develop. Biol. Stand.,* **49**, 447–454.

13

Vaccination against Infectious Pancreatic Necrosis

M. Dorson

THE DISEASE PROBLEM

Infectious pancreatic necrosis of salmonids was first recognized as a viral disease by Wolf *et al.*[1] The virus which has been classified in the family Birnaviridae[2] is of international concern since it has been recognized in every country where salmonids are present and where viral diagnostics are practised (with, perhaps, the exception of Australia, as reviewed by Dorson[3]).

IPNV is an unenveloped icosahedral virus, 60 nm in diameter (Fig. 13.1), with bi-segmented, double-stranded RNA genome. The viral capsid basically consists of three proteins (Vp_1, Vp_2, Vp_3). A fourth protein, Vp_4, is considered to be a cleavage product of Vp_3.[2]

The virus was first isolated from Brook trout (*Salvelinus fontinalis*) suffering from a severe disease with the following clinical signs: loss of equilibrium with whirling and corkscrew swimming followed by death, and a characteristically swollen belly. The name of the disease refers to the severe necrosis of exocrine pancreas revealed by histopathological examination. Many other species of the family Salmonidae are also susceptible: rainbow trout (*Salmo gairdneri*), brown trout (*Salmo trutta*), cut-throat trout (*Salmo clarki*), amago salmon (*Oncorhynchus rhodurus*), sockeye salmon (*Oncorhynchus nerka*),[3,4] and Arctic char (*Salvelinus alpinus*) (Dorson, unpublished).

Despite several records of disease presumably caused by IPNV in 1 year old fish, IPN has been considered from the beginning as a disease of fry, a facet which has been assessed experimentally.[5,6] In rainbow trout, the limit has been fixed around 1500 degrees × days post-hatch. The role of temperature is important in the development of the disease,[5,6] and IPN can be considered as a coldwater disease, occurring mainly at temperatures below 16°C but this may not be true for all IPNV strains. Independently of temperature, 'seasonal' differences in the severity of the disease have been suspected but no clear role of photoperiod has yet been implicated. The physico-chemical quality of the water has no apparent influence on the susceptibility of fish to IPN since the disease occurs in waters providing optimal conditions for the well-being of salmonids.

FISH VACCINATION
ISBN 0-12-237485-1

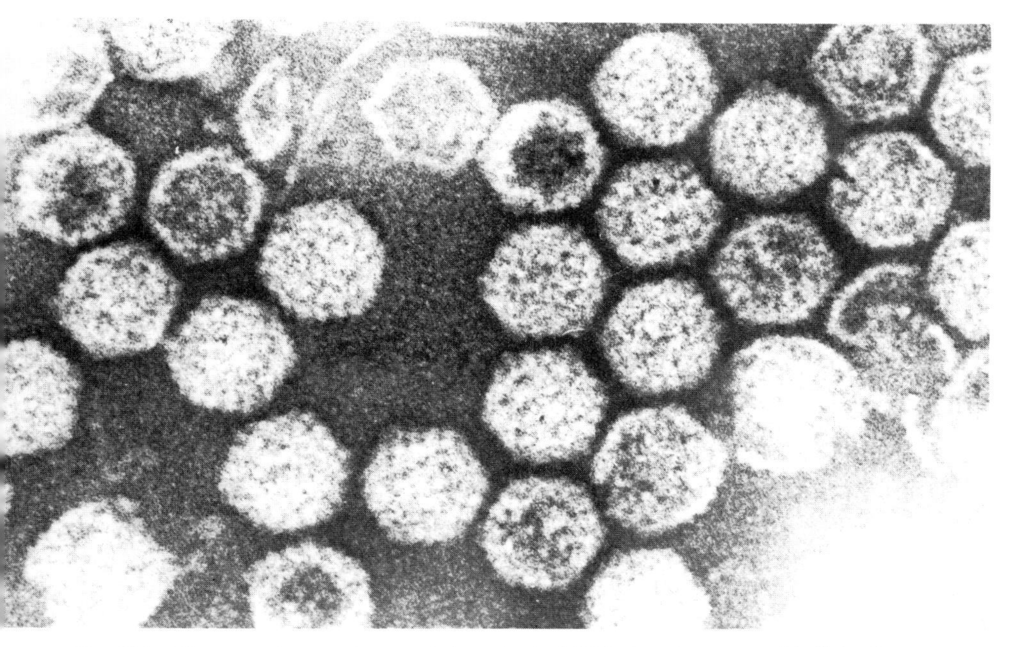

Fig. 13.1 Electron micrograph, negatively stained of IPN virus (courtesy of J. Cohen).

It has been recognized for a long time that IPN survivors can become lifelong carriers while producing circulating antibodies[7] and can shed virus with the faeces.[8] Therefore, lateral transmission can occur if susceptible fry are placed downstream of infected adults or if the virus is transmitted by fomites. Feral salmonids can act as reservoirs of the virus.[9,10] IPNV can also be frequently harboured by non-salmonid species of which a long list exists,[4] and is extending constantly. Of these species only a few have proved susceptible to IPN disease, e.g. pike,[11,12] but closely related viruses, which are not pathogenic for salmonids, cause diseases in non-salmonids e.g. EVE (eel virus European) in eel,[13] or have only been isolated in various invertebrates.[4,14]

So called vertical transmission of IPNV was long suspected from epidemiological observations and transmission of IPNV inside eyed eggs was deduced when transmission to fry was not eliminated by external disinfection of eggs from infected brood stock.[15] Experimental trials have shown that the virus can adsorb tightly to the egg shell and persist until hatching.[16] Furthermore, the virus can adsorb to the spermatozoan and penetrate the egg at fertilization. Successful transmission has been achieved with sperm incubated with viral particles before insemination.[17] It has been observed several times that IPN was not transmitted in groups of more than 10 000 eggs originating from farmed stocks known to contain carrier individuals and laboratory investigations into the presence of virus in eggs from such infected brood stocks have clearly shown that only a few eggs (less than 1 per cent) were contaminated. This can be related to the low prevalence of sires carrying virus in their milt and also explains why mortalities caused by IPNV usually become obvious around the time fry start to feed, since several cycles of infection are necessary before the whole population becomes infected.

Although conclusive experimental evidence is lacking, it is likely that disinfection of gametes with 25 ppm available iodine at the time of fertilization (as recommended by Klontz, unpublished) is an efficacious means of preventing IPNV 'vertical' transmission.

IPNV heterogeneity was first recognized[18] from differences in the degree of pathogenicity associated with different strains. Further work has confirmed this fact with the isolation of a number of strains differing by their pathogenic or antigenic properties. Moreover, it has been recognized that these properties could be easily changed by culture conditions.[19,20] When considering this heterogeneity, a first prerequisite is to limit the subject to viruses able to cause typical IPN disease in salmonids including isolates causing a mild disease, or attenuated variants obtained from pathogenic strains. This criterion excludes many 'IPN like' viruses isolated from non-salmonids and invertebrates. Even within this framework, the situation is quite complex. The first strain isolated by Wolf, typed as ATCC VR 299, was used to raise the first antisera in rabbits, but Wolf *et al.*[21] noticed that 28 strains were far from being neutralized to the same extent by this antiserum. Vestergard-Jorgensen and Kehlet[22] have clearly separated two European serotypes (Sp, which is usually highly virulent, and Ab, which displays a low virulence) from the American strain VR 299. If it has been possible, up to now, to place all European IPNV isolates into the Sp or Ab serotypes, other strains (e.g. Canada 1) have been isolated in North America which differ greatly from VR 299.[23,24] More recently another isolate, 'West Buxton', has been chosen as representative of North American strains. A panel of five monoclonal antibodies produced against this isolate has allowed Caswell-Reno *et al.*[25] to demonstrate that 14 different isolates representing four serotypes when tested with conventional rabbit antisera, could be separated into at least nine antigenically distinct viruses. This result suggests that with an increasing number of monoclonal antibodies a concurrently increasing number of isolates can be identified.

Controlling IPN disease in trout farms can be attempted by several means. Clearing the site followed by disinfection can be considered only if there is no risk of virus entering the farm in the water supply, i.e. in the absence of infected trout farms and carriers upstream. Such conditions are rare, because a large number of fish species are known to be capable of harbouring IPNV. It is a relatively common situation for the river water to be contaminated, and for carriers to be present amongst the brood stock. However, there should be no risk with spring or well water which supplies a hatchery. In this case, the disinfection of gametes with iodine, as mentioned above, would allow virus-free fry to be reared in isolation until they reached an age when risk of infection is low. Then they could be transferred to an infected area of the farm without danger. This period can be shortened and the risks of disease decreased if fry are kept at higher temperature (15–16°C) in recirculating units.

Several attempts to control IPN by chemotherapy have been published. In 1972, Economon[26] described experiments where brook trout fry treated *per os* with polyvinyl-pyrrolidone iodine were slightly more resistant to IPNV. Savan and Dobos[27] obtained a similar result with virazole. In these two cases, the differences between controls and treated fish were not sufficient to foster field application. Similarly, Elliot and Amend[28] noted that acriflavine had an *in vitro* activity against IPNV though no *in vivo* tests with this drug have been published.

Some salmonid species are clearly recognized as IPN resistant, e.g. coho salmon (*Oncorhynchus kisutch*). Hybrids obtained by inseminating rainbow trout ova with coho salmon milt while being resistant to viral haemorrhagic septicaemia, are unfortunately still susceptible to IPN.[29]

The selection of resistant salmonid strains for use in fish farming is a reasonable goal, but no successful results have been published, and the high mortalities which are still recorded in fry from populations submitted to IPN pressure over numerous generations, are not encouraging. Despite this pessimistic observation, it seems that selection of IPN-resistant rainbow trout strains is not utopian, since one has been obtained by Japanese farmers (Okamoto, pers. comm.).

THE IDEAL VACCINE AND ITS STRATEGIC USE

Because of the epidemiological characteristics of IPN, an anti-IPN vaccine should:

protect the fish very early in its life,
allow a rapid onset of the protection,
be delivered easily i.e. orally, or best via the water before fry start feeding,
protect against a wide variety of antigenically different strains in view of a world wide use,

and possess the other classical properties required for any vaccine (efficacy – innocuousness).

These prerequisites indicate a single possible answer – a non-reverting attenuated live vaccine, with the property of interfering strongly and immediately after its delivery with wild IPNV strains even before eliciting an immune response. The duration of immunity is of lesser importance since it is only young fry which are at risk from IPN-induced mortality.

OVERVIEW OF RESEARCH TOWARDS IPN VACCINATION

It must be recalled that when fish pathologists began working on IPN vaccination, there was a high degree of optimism spurred on by the promising results obtained in the field of bacterial vaccination, and also by the first positive results published for the fish rhabdoviruses – infectious haematopoietic necrosis (IHN) virus[30] and viral haemorrhagic septicaemia (VHS) virus.[31] But until that time little information was available concerning the possibility of immunizing young salmonid fry and work by Ellis,[32] which described the ontogenic appearance of lymphoid organs in the Atlantic salmon and the lack of lymphocyte membrane immunoglobulin until the time of first feeding, suggested that effective vaccination of fish prior to this stage may not be possible.

Experimental attempts to vaccinate fish against IPN have used either inactivated or attenuated virus.

Inactivated IPN vaccines

The first preliminary results[33] indicated that when rainbow trout fry started feeding (250 degrees × days after hatching), one injection of formalin-inactivated IPNV elicited total protection to experimental challenge given 2 weeks later. Oral immunization (commercial feed moistened with formalin-inactivated virus given over 3 days) did not confer any protection. Neither did the use of hyperosmotic infiltration (HI, 2 min in 53 g/litre NaCl followed by 30 min immersion in a concentrated vaccine suspension) where the mortality in treated groups equalled the mortality in controls (Table 13.1). Nevertheless, the fact that fry could be immunized by injection was encouraging and further research was undertaken in different laboratories, mainly collaboratively in Fish Diseases Laboratory, Weymouth, England, and Laboratoire d'Ichtyopathologie, Grignon, France.

The use of inactivated vaccines was explored further. Since the high concentration of formalin used for virus inactivation required neutralization or a prolonged dialysis to remove toxicity, β-propiolactone (BPL), which self-hydrolyses to non-toxic residues, was employed successfully for virus inactivation.

Injection of BPL-inactivated virus into fry led to a satisfactory protection,[34] despite the fact (as was noticed later) that BPL-inactivated virus had lost some of its antigenicity compared to formalin inactivated virus.[35] Moreover, a significant protection was recorded (19 per cent mortality in 'vaccinated' fish compared to a mortality of 53 per cent in the non-vaccinated control group) following the first vaccination trial by applying BPL-inactivated virus by the HI bath method. Unfortunately, subsequent experiments failed to confirm this promising result. In order to use the most intact (antigenically) inactivated virus, IPNV was UV-inactivated under a germicide lamp. Once again, vaccination by the HI bath method failed to provide protection (Dorson, unpublished). Formalin inactivated virus was also employed by HI bath combined with a commercial *Vibrio* vaccine (Hivax Tavolek) used as a possible adjuvant but once again without success.

Table 13.1 *Total mortalities of rainbow trout fry challenged with IPNV 2 weeks after vaccination* (from Dorson[33])

Group	Mortality %
Non vaccinated control	75
Formalin inactivated virus injected i.p.	12
Formalin inactivated virus *per os*	70
Formalin inactivated virus HI bath	77
Bath in live attenuated virus	76

A last attempt was made using virus disrupted into its constituent polypeptides by two different methods (sodium dodecyl sulphate, urea and acetic acid at 37°C or sodium dodecyl sulphate and mercaptoethanol at 100°C) with the hope that small polypeptides (molecular weight less than 100 000 daltons) would penetrate more easily and reach the immunocompetent cells. Unfortunately the results were negative again by the HI bath method (Table 13.2), and also following i.p. injection of the fragments. This last result indicated that the virus constituent proteins had lost their antigenicity following the disruption procedure. This fact was confirmed by using a rabbit antiserum against IPNV.[34]

Attenuated IPN vaccines

The field of attenuated vaccines was also investigated actively during the same period in two directions: attenuation of a wild strain under culture conditions, and a search for natural non-pathogenic strains.

Table 13.2 *Total mortalities of rainbow trout fry challenged with IPNV 2 weeks after vaccination* (from Hill *et al.*[34]).

Group	Mortality %
Non-vaccinated control	78
Sham vaccinated (cell extract injected)	50
Formalin inactivated virus injected i.p	7
Disrupted virus injected i.p.	65
Disrupted virus HI bath	54

Initially there were only two possible candidates for a live vaccine – Ab type virus, known to cause low (but significant) mortalities in the field and an attenuated variant obtained by successive passages in rainbow trout gonad (RTG) cell culture.[19] Both strains were used for early infection of fry, but this treatment did not confer any protection (Table 13.1).

The attenuated virus, obtained after several passages under routine conditions in RTG cell culture displayed an unusual property by being markedly neutralized by a protein ('6 S factor') in normal trout serum (NTS). Furthermore, this feature was found to be readily acquired by numerous strains of IPNV in trout or carp cell culture. The inability of such viruses to confer protection may have resulted from poor immunogenicity related to a poor viral growth *in vivo* as a result of their sensitivity to NTS. Serial passages were therefore performed in the presence of NTS. Surprisingly the yields of infectious virus declined progressively with increased passage, as if growth *in vitro* and resistance to NTS were incompatible.[34] Despite this snag, different attenuated strains were obtained (including one obtained at 20°C) but as soon as a virus had lost pathogenicity for fry, it appeared to lose immunogenicity.

Table 13.3 *Total mortalities of rainbow trout fry challenged with IPNV 2 weeks after vaccination with live avirulent strain 74/53. Three different experiments (a, b, c).* (From Hill *et al.*[34], and unpublished.)

Group		Mortality %
Non-vaccinated control	(a)	45
Simple bath method	(a)	16
Non-vaccinated control	(b)	53
HI bath method	(b)	18
Non-vaccinated control	(c)	67
Simple bath method	(c)	55
HI bath method	(c)	32

Finally, interesting results were reproducibly obtained only with an avirulent strain (strain 74/53) isolated from perch (*Perca fluviatilis*) in England.[36] Surprisingly, better results (i.e. protection indices following challenge 2 weeks later) were constantly obtained when this 'live' vaccine was delivered by the HI method. Nevertheless the protection was never complete, and fluctuated between 25 and 75 per cent depending on the experiments (Table 13.3).

CONCLUSIONS AND PROSPECTS

To date no IPN vaccine has ever been marketed and the author is unaware of any field trials being performed. Despite the theoretical possibility of vaccinating fry, it is obvious that any inactivated vaccine (including a recombinant immunogenic protein) would remain too expensive compared to the value of fish and to the shortness of the period during which fry are susceptible. One may consider the usefulness of vaccinating broodstock by injection. Such immunization prevents intravenous or water-borne infection of adults[12,37] but it is unlikely to eliminate existing virus carriers since it is well known that trout can remain carriers despite high circulating antibody levels. Furthermore, there is no evidence of any transfer of immunity from mother to fry,[12] and therefore, no benefit would result from mother immunization. Attenuated vaccines appear especially suitable to the case of IPN, as explained above. Unfortunately an appropriate strain has not yet been found and efforts to obtain the 'wonder' strain have currently ceased. Workers are also dissuaded by the general fears of live vaccines reverting to virulent types and judging by the apparent instability of every IPNV strain so far studied such fears may be justified.

REFERENCES

1 Wolf, K., Snieszko, S. F., Dunbar, C. E. and Pyle, E. (1960). Virus nature of infectious pancreatic necrosis in trout. *Proc. Soc. Exper. Biol. Med.*, **104**, 105–108.

2 Dobos, P., Hill, B. J., Hallett, R., Kells, D. T., Becht, H. and Teninges, D. (1979). Biophysical and biochemical characterization of five animal viruses with bisegmented double-stranded RNA genomes. *J. Virol.* **32**, (2), 593–605.

3 Dorson, M. (1983). Infectious pancreatic necrosis of salmonids. Overview of current problems. In: *Antigens of Fish Pathogens*, ed. Anderson, Dorson and Dubourget, Collection Fondation Marcel Mérieux, pp. 7–31.

4 Hill, B. J. (1982). Infectious pancreatic necrosis virus and its virulence. In: *Microbial Diseases of Fish*, ed. R. J. Roberts, London: Academic Press, pp. 91–114.

5 Frantsi, C. and Savan, M. (1971). Infectious pancreatic necrosis virus. Temperature and age factors in mortality. *J. Wildl. Dis.*, **7**, 249–255.
6 Dorson, M. and Torchy, C. (1981). The influence of fish age and water temperature on mortalities of rainbow trout (*Salmo gairdneri* Richardson) caused by an European strain of infectious pancreatic necrosis virus. *J. Fish Dis.*, **4**, 213–221.
7 Wolf, K., Quimby, M. C. and Bradford, A. D. (1963). Egg-associated transmission of IPN virus of trouts. *Virology,* **21** (3), 317–321.
8 Frantsi, C. and Savan, M. (1971). Infectious pancreatic necrosis virus: comparative frequencies of isolation from feces and organs of brook trout (*Salvelinus fontinalis*). *J. Fish. Res. Board Can.*, **28** (7), 1064–1065.
9 Yamamoto, T. and Kilistoff, J. (1979). Infectious pancreatic necrosis virus. Quantification of carriers in lake populations during a 6-year period. *J. Fish. Res. Board Can.*, **36**, 562–567.
10 Munro, A. L. S., Liversidge, J. and Elson, K. G. R. (1976). The distribution and prevalence of infectious pancreatic necrosis virus in wild fish in Loch Awe. *Proceedings of the Royal Society of Edinburgh*, Section B (Natural Environment), **75**, (4), 223–232.
11 Ahne, W. (1980). Occurrence of infectious pancreatic necrosis virus (IPN) in different fish species. *Berl. Munch. Tierarztl. Wochenschr.*, **93**, (1), 14–16.
12 Dorson, M. (1985). La nécrose pancréatique infectieuse des Salmonidés: étude de la maladie, du virus et de la réaction de la truite. Thèse présentée à l'Université Paris VII.
13 Sano, T., Okamoto, N. and Nishimura, T. (1981). A new viral epizootic of *Anguilla japonica* Temminck and Schlegel. *J. Fish Dis.*, **4** (2), 127–139.
14 Hill B. J. and Way, K. (1980). Properties and interrelationship of bisegmented double-stranded RNA viruses of fish and shellfish. Proceedings of the conference on aquatic animal viruses. Institut Pasteur, pp. 22–24.
15 Bullock, G. L., Rucker, R. R., Amend, D., Wolf, K. and Stuckey, H. M. (1976). Infectious pancreatic necrosis: transmission with iodine treated and non treated eggs of brook trout (*Salvelinus fontinalis*). *J. Fish. Res. Board Can.*, **33** (5), 1197–1198.
16 Ahne, W. and Negele, R. D. (1985). Studies on the transmission of infectious pancreatic necrosis virus via eyed eggs and sexual products of salmonid fish. In: *Fish and Shellfish Pathology,* ed. A. E. Ellis, London: Academic Press, pp. 261–269.
17 Dorson, M. and Torchy, C. (1985). Experimental transmission of infectious pancreatic necrosis virus via the sexual products. In: *Fish and Shellfish Pathology,* ed. A. E. Ellis, London: Academic Press, pp. 251–260.
18 Wolf, K. (1972). Advances in fish virology: a review 1966–1971. *Symp. Zool. Soc. Lond.,* **30**, 305–331.
19 Dorson, M., Castric, J. and Torchy, C. (1978). Infectious pancreatic necrosis virus of salmonids: biological and antigenic features of a pathogenic strain and of a non-pathogenic variant selected in RTG2 cells. *J. Fish Dis.,* **1**, 309–320.
20 Nicholson, B. L., Thorne, G. W. and Janicki, C. (1979). Studies on a host range variant from different isolates of infectious pancreatic necrosis virus. *J. Fish Dis.,* **2**, 367–379.
21 Wolf, F., Bullock, G. L., Dunbar, C. E. and Quimby, M. C. (1968). Viral diseases of freshwater fishes and other lower vertebrates: comparative studies on IPN virus. *Progr. Sport. Fish. Res.,* **77**, 138–140.

22 Vestergard-Jorgensen, P. E. and Kehlet, N. P. (1971). Infectious pancreatic necrosis (IPN) viruses in Danish rainbow trout: their serological and pathogenic properties. *Nord. Vet. Med.* **23**, 568–575.
23 Nicholson, B. L. and Pochebit, S. (1981). Antigenic analysis of infectious pancreatic necrosis viruses (IPNV) by neutralization kinetics. *Develop. Biol. Standard.*, **49**, 35–41.
24 MacDonald, R. D., Moore, A. R. and Souter, B. W. (1983). Three new strains of infectious pancreatic necrosis virus isolated in Canada. *Can. J. Microbiol.*, **39**, 137–141.
25 Caswell-Reno, P., Reno, P. W. and Nicholson, B. L. (1986). Monoclonal antibodies to infectious pancreatic necrosis virus: analysis of epitopes and comparison of different isolates. *J. Gen. Virol.*, **67**, 2193–2205.
26 Economon, P. P. (1972). Polyvinyl pyrrolidone-iodine as a control for infectious pancreatic necrosis of brook trout. *F.I. EIFAC 72/SC II.* Symp. 13.
27 Savan, M. and Dobos, P. (1980). Effect of virazole on rainbow trout *Salmo gairdneri* Richardson fry infected with infectious pancreatic necrosis virus. *J. Fish Dis.*, **3** (5), 437–440.
28 Elliot, D. G. and Amend, D. F. (1978). Efficacy of certain disinfectants against infectious pancreatic necrosis virus. *J. Fish. Biol.*, **12**, 277–286.
29 Dorson, M. and Chevassus, P. (1985). Etude de la réceptivité d'hybrides triploïdes truite arc-en-ciel x saumon coho à la nécrose pancréatique infectieuse et à la septicémie hémorragique virale. *Bull. Fr. Pêche Pisci.*, **296**, 29–34.
30 Amend, D. F. (1976). Prevention and control of viral diseases of salmonids. *J. Fish. Res. Board Can.*, **33**, 1059–1066.
31 De Kinkelin, P. and Le Berre, M. (1977). Démonstration de la protection de la truite arc-en-ciel contre la SHV par l'administration d'un virus inactivé. *Bull. Off. int. Epiz.*, **87** (5–6), 401–402.
32 Ellis, A. E. (1977). Ontogeny of the immune response in *Salmo salar*. Histogenesis of the lymphoid organs and appearance of membrane immunoglobulin and mixed leucocyte reaction. In: *Developmental Immunobiology*, eds. J. B. Solomon and J. D. Horton, Amsterdam: Elsevier/North Holland Biomedical Press, pp. 225–231.
33 Dorson, M. (1977). Vaccination trials of rainbow trout fry against infectious pancreatic necrosis. *Bull. Off. int. Epiz.*, **87** (5–6), 405–406.
34 Hill, B. J., Dorson, M. and Dixon, P. F. (1980). Studies on immunization of trout against IPN. In *Fish Diseases*. Third COPRAQ-Session, ed. W. Ahne, Berlin: Springer-Verlag.
35 Dixon, P. F. and Hill, B. J. (1983). Inactivation of infectious pancreatic necrosis virus for vaccine use. *J. Fish Dis.* **6**, 399–409.
36 Hill, B. J. and Dixon, P. F. (1977). Studies on IPN virulence and immunization. *Bull. Off. int. Epiz.*, **87** (5–6), 425–427.
37 Sano, T., Tanaka, K. and Fukuzaki (1981). Immune response in adult trout against formalin killed concentrated IPNV. *Develop. Biol. Stand.* **49**, 63–70.

14

Vaccination against Viral Haemorrhagic Septicaemia

P. de Kinkelin

Despite the investigations which have been carried out during the last 15 years in Denmark and France there is no licensed vaccine against VHS and this paper aims to explain why.

Viral haemorrhagic septicaemia (VHS) is a coldwater rhabdovirus infection of trout, grayling, white fish and pike occurring in continental Europe but is mainly known because of its clinical and economic consequences in trout farming. Infection usually results in death, due to the impairment of the salt–water balance which occurs in a clinical context of oedema and haemorrhages. Virus multiplication in endothelial cells of blood capillaries, leucocytes, haematopoietic tissue and nephron cells, underlies the clinical signs (Figs. 14.1–14.3).

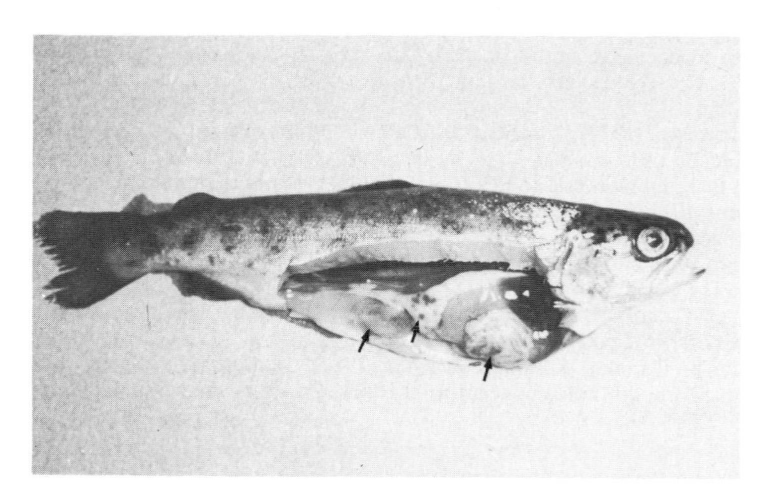

Fig. 14.1 Acute form of viral haemorrhagic septicaemia (VHS) in rainbow trout. Arrows indicate the haemorrhages.

FISH VACCINATION
ISBN 0-12-237485-1

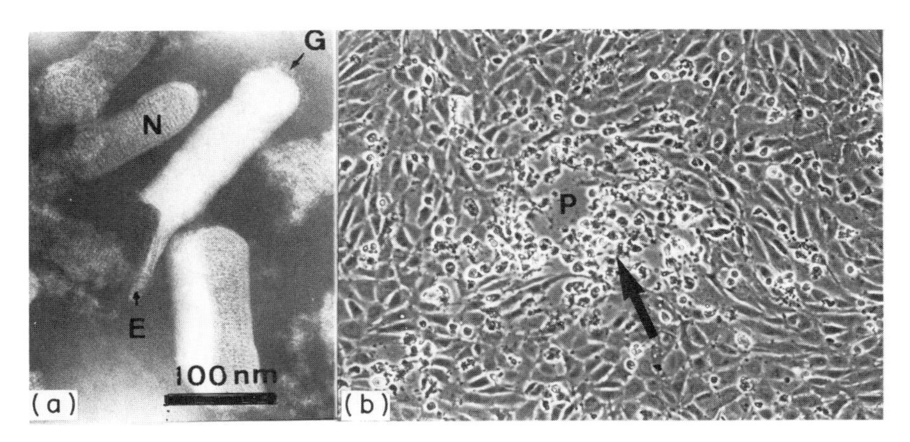

Fig. 14.2 (a) Negatively stained viral particles of VHS virus showing their helicoidal nucleocapsid (N), spikes (G) and envelope (E), trespassing the flat end of the particle. (Courtesy J. Cohen). (b) Initial cytopathic effect induced by VHSV in a monolayer of the trout cell line RTG 2. The infected cells become spherical (arrow) and fall away from the plastic surface forming a plaque. This illustrates the mode of production of VHSV and thus of the vaccine.

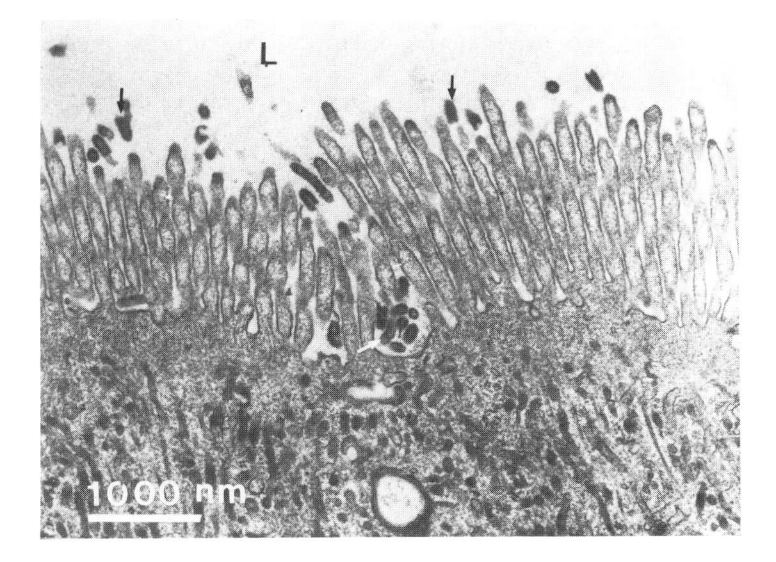

Fig. 14.3 Ultrathin section of a kidney tubule cell of a VHSV infected trout (3 days post-infection) showing the first steps of virus shedding; the viral particles (arrows) originating from the infected cell enter the lumen (L) of the tubule and hence the urine and water. (Courtesy S. Chilmonczyk).

THE DISEASE PROBLEM

A tentative assessment of the economic significance of VHS is about 30 million pounds per year,[1] on the basis of information supplied from several European fish disease laboratories. Epidemiology provides a pragmatic approach to an understanding of the VHS problem.

Epidemiological facets of VHS

Source and transmission of virus

The sources of the virus are clinically infected fish and asymptomatic carriers from both cultured and feral fish. For many years rainbow trout was considered to be the only susceptible species but the virus has now been isolated from a wide variety of other species (see below).

Virulent virus is shed in the faeces, urine, sexual fluids while kidney, spleen, brain and digestive tract are the sites in which the virus is the most abundant.

Contamination is horizontal and may be direct or through a vector, water being the major abiotic vector. The fish farmer and fomites are vectors if precautions are not taken. Birds have been involved[2] but their role seems to be restricted to localities neighbouring fish farms. The surface of the egg may be contaminated in infected spawners but the virus is easily removed by the water flow long before hatching and is readily controlled by disinfection of eggs with iodine.

As with many other pathogens shed into the water by diseased or asymptomatic carrier fish, VHS virus can spread into trout farms located along the water flow.

As long as water systems are kept separate from VHS infected fish, the spread of the disease is avoided. For example, although VHS is endemic in many geographic areas of France, Brittany remains free of VHS because its river basins are not connected with those of other regions and because the local trout farmers do not import live fish into Brittany. Each time a trout farmer contravened that rule resulting in a VHS outbreak, an immediate slaughter of infected and suspected stocks has prevented further spread of the disease despite the high density of the farms in some Brittany districts.

Conversely, once VHS virus is established in a farm stock and therefore in the water shed, the disease becomes endemic because of carrier fish.[3] The presence of carrier fish is ineluctable in the usual context of trout farming which aims to supply marketable size live fish throughout the year for consumption and/or angling. Under these circumstances, fish production by successive waves followed by disinfection which could break the cycle of infection is not possible. Thus, young fish originating from hatcheries, which are usually VHS free since their water supplies come from spring or well water, are under constant threat from carrier fish.

Host and virus factors influencing susceptibility

Host factors and susceptibility

Nine species are presently known to be susceptible to VHS infection leading to clinical diseases: rainbow trout (*Salmo gairdneri*), brown trout (*S. trutta*),[4] lake trout (*Salvelinus namaycush*) (Dorson, unpublished), pike (*Esox lucius*),[5] grayling (*Thymallus thymallus*),[6] white fish (*Coregonus* sp.),[7] sea bass (*Dicentrarchus labrax*) and turbot (*Scophthalmus maximus*),[8] and sea bream (*Chrysophris aurata*) (Castric, unpublished). Susceptibility of trout, pike, grayling and white fish was observed under natural conditions whereas that of the other species cited was established experimentally by water-borne infection and indicates that these species are potential targets for the disease. Differences in resistance to VHS amongst salmonid species has been used as an approach to genetic control of VHS through intergeneric hybridization (see below).

There is a high degree of individual variability within a species to susceptibility to VHS and this causes problems to the reproducibility of vaccine testing[9] and constitutes a major problem for the use of live vaccine because the susceptibility of fish to infection by an attenuated virus strain is made difficult to predict and thus hinders assessment of the vaccine's safety. However, this variability, if shown to be based upon inheritable factors governing the degree of resistance, may allow selective breeding programmes to improve resistance in farmed stocks.

The age of fish seems important – the younger the fish, the higher the susceptibility to clinical disease. The influence of sex is unknown. Concerning the effect of environmental parameters, VHS is unusual in that poor water quality e.g. low oxygen, high nitrogen wastes,[10] and malnutrition reduce susceptibility to VHS. Furthermore, factors detrimental to the trout immune system (e.g. x-irradiation) do not enhance susceptibility to VHS but appear to decrease the time to death.[11]

Similarly, intercurrent infections such as infectious pancreatic necrosis, furunculosis, bacterial infections of gills and ichthyophthiriosis decrease the susceptibility of trout to VHS. However, we have no information about the effects of poor environmental conditions on virus shedding by carrier fish.

The most prominent environmental factor affecting VHS is unquestionably water temperature. Clinical disease does not occur above 14°C under natural conditions. There is evidence that the kinetics of the host natural immunity, and especially interferon synthesis,[12] is responsible for this resistance since interferon synthesis is temperature dependent and has been shown to play a protective role against VHS.[13]

Genetic factors in the virus

Presently, five serotypes of VHSV are recognized. The virulence of virus strains appears to be correlated with the virus serotype. Strain 23/75[14] is pathogenic both for rainbow and brown trout and is the most virulent VHS strain, killing with a fair reproducibility between 90 and 100 per cent rainbow trout of 1500°C days following water-borne infection at 10°C. Strain 23/75 belongs to the third serotype of VHS.[15] The French strain 07/71 and the Danish F_1 (both type 1 strains) are only pathogenic for rainbow trout and exhibit 'standard' virulence since they kill about 80 per cent of fish tested under similar conditions as above. Mortalities due to strain He (type 2)[29] seldom exceeds 60 per cent. Distinct from these three types of VHSV, two others which are not neutralized by the existing antisera have been isolated during the past 5 years. They are isolate 02/84 from rainbow trout in our laboratory and isolate 49/82 from rainbow trout in FRG,[16] these two isolates being serologically different.

Despite the serological heterogeneity which is also detectable with trout antisera, passive transfer of antibodies to type 1 induced protection against challenge by all three virus types.[17] Similarly, fish immunized with either killed or live vaccine showed a certain degree of cross-protection.

Beside the theoretical possibility of vaccination, there are other approaches to VHS control and one of these has been successfully used in Denmark for more than 20 years.

Control by means other than vaccination

Official programmes of fish health control

Such programmes are based upon policy coordinated on a national scale, or at least on a water-shed scale. The general methodology involves virus certification of fish stocks, controlled movement of fish stock, destruction of infected populations, disinfection of premises and restocking the farms with healthy fish.[18] The policy requires human, financial and legal resources and, of course, the close collaboration of the fish farmers.

In Denmark, a VHS control programme was begun in 1965 on a voluntary basis and was followed by legislation in 1969. This programme which also prohibited the restocking of rivers for angling purposes resulted in eradication of the disease in more than 350 farms within a 12 year period in an area of about 10 000 km^2.

Obviously, a policy such as this will only work with the full cooperation of fish farmers. Fish health control for VHS remains of prime importance each time a trout farm is established on a virgin water-shed or in a country which plans to start a trout farming industry. At the international level basic information for fish health control is included in the International Animal Health Code of the Office International des Epizooties, Paris.

Genetic approach of VHS control

Intergeneric hybridization between male coho salmon (*Oncorhynchus kisutch*) and female rainbow trout resulted in a low production and survival rate of hybrids but these fish proved to be resistant to VHS. Triploidization of the hybrid eggs, induced by heat shock after fertilization of ova, led to triploid progeny of fair viability (around 60 per cent). This progeny was far less susceptible to VHS than the diploid and triploid rainbow trout.[19] Identical results were obtained with triploid hybrids from brook trout (♂) (*Salvelinus fontinalis*) and rainbow trout. Field observations confirmed the laboratory infection trials and triploid hybrids seem to constitute an immediate, efficient and promising method to avoid VHS in heavily infected areas. From a zootechnical view point, brook trout × rainbow trout are the most promising since the sires are available in Europe and the progeny have an attractive appearance and a fair growth rate.

In parallel, experiments have shown that the progeny of certain rainbow trout sires was significantly less susceptible to VHS than others (Dorson, unpublished). Such selective breeding programes have only recently begun but the encouraging results so far may lead to an efficient method of control of VHS.

THE IDEAL VACCINE AND ITS STRATEGIC USE

The ideal vaccine against VHS should be (a) potent and provide protection lasting for the whole economic life of the fish independently of being boosted by natural infection; (b) safe; (c) polyvalent; (d) cheap (around 0.2–0.3 UK pence per fish aged 1500°C days); (e) easy to deliver (from fish farmer's view point the best would be oral); (f) stable, thus remaining potent during storage and shipping; (g) capable of promoting effective resistance to the disease but not necessarily to the infection; and (h) should not hinder the diagnostic techniques used in epidemiological surveillance of control policies. Strategically speaking, the vaccine should be administered to young fish prior to the possibility of them being exposed to virus i.e. in infected areas, when they are still in the hatchery. Since vaccination must be performed by a mass delivery method, the minimum requirements for the vaccine are potency and safety. Unfortunately, in practice one is confronted with the dilemma 'safety or potency'? None of the experimental vaccines tested fulfill all these requirements.

NATURE OF THE VACCINE

As with all vaccines against viruses, those against VHS are either killed or live vaccines. Their respective characteristics are given in Table 14.1.

Killed vaccine must be injected as no immunity was induced following immersion vaccination[16] presumably because the inactivated viral particles do not penetrate, at least in a sufficient amount, into the fish. Thus its use is limited to high value fish, such as those which are intended for seawater rearing or to countries which have different notions of cost-effectiveness from occidental ones. On the other hand, live vaccine can be delivered via the water route and since the virus particles multiply in the host they do not need to be introduced in large amount. Unfortunately, live vaccines are not completely innocuous and are most often banned by legislation (see below).

VACCINE TRIALS AND VACCINATION EFFECTIVENESS

The methods for assessing potency and economic efficacy of fish vaccines, including the anti-VHS vaccines, have been critically reviewed by Michel *et al.*[9]

Laboratory vaccination trials

Because of the difficulties in testing vaccines, due to the variability in susceptibility of fish and the 'group effects' encountered in small-scale experiments, the method employed by our laboratory to assess vaccine potency involves conducting trials in duplicate or triplicate aquaria with a minimum number of 30 fish per aquarium. The tests were performed with different batches of both killed and live vaccine delivered to fish from different origins during various seasons of the year. It is assumed that if a vaccine is potent it necessarily induces protection in laboratory tests though it may fail or provide less satisfactory results when tested in the field. However, if it fails to induce protection in laboratory trials it probably never would succeed in the field.

Table 14.1 *Characteristics of the anti-VHS experimental vaccines*

Type of vaccine / Characteristics	Killed vaccine[1]	Live vaccines		
		REVA[2]	F25(21)[3]	07.71 variants[4]
Virus strain	07.71 or any pathogenic strain	F1	F1	07.71
Cell line	EPC[5]	RTG[6]	EPC	EPC
Mode of production	Inactivation by β-propiolactone 1 : 5000/24 h at 15–20°C and addition of formalin 1 : 2000	240 successive sub-cultures in RTG cells, at 14°C	Following about 100 sub-cultures in RTG cells, 81 passages in EPC at progressively increasing temperatures up to 25°C (21 passages at 25°C)	According to the variant 100 sub-cultures either at 14°C or at progressively increasing temperatures up to 25°C (43 passages at 25°C)
Storage	+4°C (for more than 2 years) or −20°C	−20°C or below	−20°C or better −70°C	−20°C or better −70°C
Instructions of use	At 100°C: i.p injection with 2×10^6 "pfu equiva,ent"/g body weight at least 30 days before challenge (delay shortened or lengthened depending on temperature)	Immersion in 1×10^4 pfu/ml water, 1 h. Optimum temperature for potency: below 10°. Intended for fish up to 100 g	Immersion into 5×10^5 pfu/ml water, 15 min (lengthen the time with lower vaccine concentrations). Optimum temperature for potency; below 10°C intended for fish 1500–2000°C days	As for F25(21)
Range of protection without booster	From 30 to 100 days at 10°C	120–150 days below 10°C below 10°C	From 1 to 50 days at 10°C	From 1 to 100 days at 10°C

(1) Ref. 17; (2) Refs. 20 and 21; (3) Refs. 23 and 24; (4) Ref. 25; (5) Ref. 27; (6) Ref. 28.

Table 14.2 *Results of immunization against VHS with killed vaccine.* (De Kinkelin, Bearzotti, Bernard, Castric, unpublished results.)

Course of trials / Trial groups	Immunization procdures			Challenge	Percentage of survivors 30 days after challenge
1700°C × days				day 60	
100 × 2 2 g fish	none	cell culture supernatant	i.p	none	95
100 × 2 2 g fish	none	cell culture supernatant	i.p.	5×10^4 pfu/ml water	28
100 × 2 2 g fish	BPL* + formalin	5×10^4 'pfu'/ml/3 h	bath	for 3 h	45
100 × 2 2 g fish	BPL* + formalin	2×10^6 'pfu'/g body weight	i.p.	for 3 h	90
4300°C × days				day 50 in sea water	
40 × 2 200 g fish	none	cell culture supernatant	i.p	none	98
40 × 2 200 g fish	none	cell culture supernatant	i.p	2×10^8 pfu/fish i.p.	50
40 × 2 200 g fish	BPL* + formalin	4×10^8 'pfu'/fish	i.p.	2×10^8 pfu/fish i.p.	86
2000°C × days				day 30	
30 × 4 5 g fish	none	cell culture supernatant	i.p.	none	95
30 × 4 5 g fish	none	cell culture supernatant	i.p.	5×10^4 pfu/ml water	30
30 × 4 5 g fish	BPL*	5×10^4 'pfu'/ml/3 h	bath	5×10^4 pfu/ml water	41
30 × 2 5 g fish	BPL*	1×10^7 'pfu'/fish	i.p.	5×10^4 pfu/ml water	78
2500°C × days				day 70	
40 5 g fish	none	saline solution	i.p.	none	97
40 5 g fish	none	saline solution	i.p.	1×10^2 pfu/fish i.m.	23
40 5 g fish	none	Freund's complete adjuvant (FCA)	i.p	1×10^2 pfu/fish i.m.	13
40 5 g fish	U.V. inactivated purified virus (5 μg/fish) + FCA		i.p.	1×10^2 pfu/fish i.m.	58

*BPL, β-propiolactone.

Killed (inactivated) vaccines

Results presented in Table 14.2 indicate that anti-VHS protection was induced following injection of the antigen but water route delivery was ineffective. Preliminary results which indicated some protection by the immersion method of delivery[4] were not confirmed and were probably due to the variability and 'group effects' (Michel *et al.*, 1984)[9] which we had not fully experienced at that time of our investigations.

Resistance to challenge was elicited in all the immunization trials regardless of fish size and it lasted for over 100 days. Moreover, the same level of protection was achieved with one i.p. injection of 2×10^6 'pfu equivalent'/g of fish as two injections of 1×10^6 'pfu' given 1 month apart and addition of 10 per cent Freund's complete adjuvant to the vaccine suspension did not enhance the protection. Lastly, a significant crossed protection between the three so-called serotypes was also demonstrated (data not shown).

The presence of virus-neutralizing antibodies in the sera of vaccinated fish was not always detected prior to challenge but in one group of trout immunized with UV-inactivated purified virus, neutralizing antibody titres correlated well with resistance to challenge.

Live vaccines

Reva strain. The production of this attenuated strain is shown in Table 14.1. The first successful vaccination trial of subadult rainbow trout was reported by Vestergaard-Jorgensen.[20] Later, more complete data clearly substantiated the potency of the Reva strain vaccine for 11 g and 100 g trout.[21] These trials also demonstrated the importance of water temperature at which fish were kept after immunization, on the efficacy (and possibly duration) of protection – the lower the temperature, the better the protection. Indeed, following immunization at 10°C, fish were kept at this temperature for 5 days and groups were then transferred to 5, 10, 15 and 20°C. The survival rates at challenge performed 9 weeks later at 10°C, were respectively 75, 40, 40 and 20 per cent. There was no survival among the non-immunized controls. At 5°C, Reva virus could be reisolated from unchallenged fish 14 weeks post immunization whereas at 15°C, virus persistence did not exceed 1 week. This suggests that interference could play a role in protection (see nature of immunity, below). Nevertheless, neutralizing antibodies were detected 12 weeks post vaccination in all the immunized series, their titres being 40–160 at 5°C and 10–80 at 20°C, which correlates well with the resulting protection.

Innocuousness of the Reva strain was fair (4–6 per cent mortality within 9 weeks) as was stability of its attenuation which was tested through 20 *in vivo* passages in young fish.

Vaccines derived from passages in cyprinid cells. Thermoresistant variant strain F25(21) was the first strain obtained in EPC cell cultures.[22] Its potency was clearly established through many trials, revealing a significant degree of protection detected from day 1 to day 45 post immunization.[23] Indeed, in all trials the survivors in immunized groups, were always at least 30 per cent higher than in non-immunized controls, following experimental and, in one case, natural challenge (Table 14.3). Protection was effective against three sero-types of VHSV. However, no neutralizing antibodies could be demonstrated suggesting that the protection was mainly due to interference and would thus last only as long as the variant virus strain continued to be present in the fish.

Table 14.3 *Results of some immunization trials against VHS with the thermoresistant variant F25 (21) administered by immersion to young rainbow trout (age: 1500°C days, weight 2 g). (After Refs. 23 and 24).*

Course of trials / Trial groups	Immunization		Challenge (1)	
	Virus strains	% of survivial 20 days post immersion	Virus strains	% of survival 28 days post challenge
2000°C × days				
73 × 2 3 g fish	none	97	07.71 (3)	27
70 × 2 3 g fish	F25 (21) (2)	93	07.71	53
73 × 2 3 g fish	none	98	23.75 (3)	2
65 × 2 3 g fish	F25 (21)	87	23.75	39
1800°C × days				
100 × 2 2.5 g fish	none	95	none	93
100 × 2 2.5 g fish	none	96	07.71 (3)	27
100 × 2 2.5 g fish	F25 (21) (2)	94	07.71	79
2000°C × days				
4300 4 g fish	none	95	(4) wild strain	45
4700 4 g fish	F25 (21) (2)	93	serotype 1	86
1000°C × days				
75 × 2 1.2 g fish	none	99	none	97
75 × 2 1.2 g fish	none	100	07.71 (3)	38
75 × 2 1.2 g fish	F25 (21) (2)	98	07.71	69
75 × 2 1.2 g fish	07.71	33	07.71	20

(1) 20 days post-immunization
(2) Immunization by immersion in 5.10^5 pfu/ml of water for 20–30 min.
(3) Experimental challenge by water route 5.10^4 pfu/ml for 3 h.
(4) Natural challenge in fish farm.

Further investigations[24] showed that neutralizing antibody did exist but that it was directed only against F25(21) variant which was antigenically rather different from the wild VHSV strain type 1. This finding provides further evidence that protection was mainly due to interference and not specific acquired immunity.

The relativity of the concept of innocuousness appears in Table 14.3 which shows that aliquots of the same F25(21) virus stock induced mortalities varying from 2 to 13 per cent in young fish of several shipments originating from the same site, as well as in fish from different origins.

The stability of the attenuation was demonstrated to be effective through serial *in vivo* passages and F25(21) appeared to be safe for the environment since clinical disease was never transmitted to healthy fish maintained in contact with vaccinated ones or to those inhabiting the ponds or raceways located downstream.

Besides the F thermoresistant strain of VHSV, some other virus strains were similarly attenuated in EPC cells. This work also demonstrated that the loss of virulence was induced by the cell and not by the adaptation to high temperature which nevertheless still constitutes a marker of the vaccine strains obtained from EPC cells.[25] Among the more recently obtained attenuated virus strains, the 07/100 (14, cloned) and the 07/100 (25) have retained their immunogenicity to the wild virus strain[26] and their potency (Fig. 14.4).

As in other trials with live vaccines the protection was never complete. However, as shown in Fig. 14.4 degrees of protection were elicited by the different vaccine strains even when delivered 24 h prior to challenge which confirmed an interference mediated protection. The final survival rates of the immunized groups ranged from 39 to 58 per cent, while survival of non-immunized controls was between 8 and 15 per cent and that of the non-challenged mock immunized group was 87 per cent. Thus, when innocuousness was fair, potency was rather low (Fig. 14.4, for vaccine 1, strain 07/14 (100 cloned)). Conversely when the vaccine was more virulent, its potency for protecting survivors against challenge was higher (vaccine 2, strain 07/100 (25)) but the final balance of survival was almost the same for both vaccine strains.

Nature of immunity

The results of experiments reported above provide evidence for protective mechanisms based upon both non-specific and specific immunity, the former being able to hinder the onset of the latter.

Early protection against challenge performed 24 h post-immunization with F25(21) virus strain or 07-71 attenuated virus strain must be due to a non-specific mechanism, namely interferon synthesis. This is likely since the protective role of trout interferon has been previously established *in vivo*.[13]

On the other hand, neutralizing antibodies are at least one of the

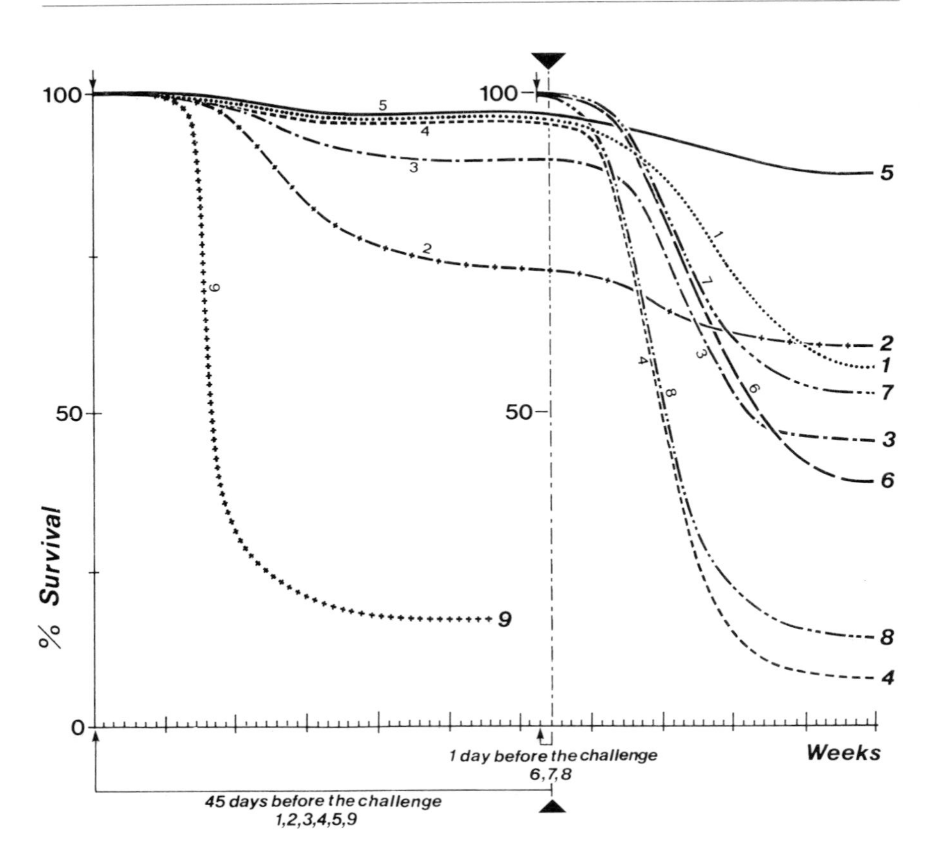

Fig. 14.4 Survival curves from vaccination trials demonstrating the resistance to VHS induced in rainbow trout vaccinated by immersion (15 min in 5×10^5 pfu/ml water) with EPC cell-adapted variants of VHSV, delivered either on day 45 or on day 1 before challenge ▲. 1 and 6: variant strain 07/14 (100) cloned (selected at 14°C); 2: variant strain 07/25 (100) (selected at 25°C); 3 and 7: variant strain F25 (21); 5: mock-immunized, non-challenged controls; 4,8: mock-immunized challenged controls; 9: wild strain 07/71 infected controls. The trials were performed at 11°C ± 1. The age of fish was 1600 °C days at the time of vaccine delivery.

components of specific immune protection as shown by the passive immunization experiments. Data on antibodies from experiments using injectable killed vaccines are scarce since the technical conditions of the seroneutralization test with trout sera[26] had not been established at the time of the experiments, but it appeared later that these neutralizing antibodies did exist (Table 14.2). In the experiments using F25(21) virus strain reported in 1981, a lack of antibodies to the virulent virus strain was reported but later it was discovered that the immunogenicity of F25(21) strain was different from that of the wild virus strain type 1.[24] Indeed, sera from anti-F25(21) hyperimmunized trout cross reacted only slightly with wild virus strain type 1 and though passively transferred F25(21) antiserum conferred significant resistance to VHS, this was less than that induced by homologous sera.

A characteristic of the antibody responses to VHS vaccines is the weakness of the viral-neutralizing titres in serum, even when measured following a 16 h test at 4°C which increases by 20 the values obtained after a 1 h neutralization test at 20°C. Since this last point was unknown at the time when immunization trials with killed vaccine and F25(21) virus strain were carried out, the antibody response of the fish was assayed by the 1-h seroneutralization test. Thus the presence of low titres of neutralizing antibodies would have been missed. Since, from the general view point of antiviral immunity, even low levels of circulating antibodies can prevent viral septicaemic infections, such an immunological status could account for the observed resistance of trout to VHS.

Reva vaccine[21] offers a good example of the association of specific and non-specific immunity in induced resistance to VHS – it appears to involve an antibody response but much of the evidence suggests that persistence of the vaccinating virus for long periods at low temperature (5°C) may improve protection by interfering with replication of the virulent virus, while at intermediate temperatures (10–15°C) the induced production of interferon may play a role. At higher temperatures (20°C) live vaccines (curiously) induced lower antibody titres than at 5°C and also the protection was lower probably because of the more rapid elimination of the vaccinating virus (possibly by a high rate of interferon production) thus reducing stimulation of the antibody response.

Thus, the Reva live vaccines may provide protection via at least two interacting processes – interference with viral replication (probably by interferon production) and antibody synthesis. For maximum protection the optimum outcome of the interaction appears to be achieved at temperatures of 5–10°C.

It is obvious that the nature of immunity against VHS is worth investigating more thoroughly since many points still remain unclear.

Field trials

Although the potency of anti-VHS vaccination was clearly established in laboratory trials, its efficacy i.e. its potency at the fish farm management level which takes into account the cost:benefit ratio, has not yet been demonstrated. To date only a small number of field trials with live vaccines have been carried out, mainly because of rather restrictive official authorizations. Moreover, because of uncontrolled environmental conditions the efficacy of the vaccination trials could not be assessed. The major snags encountered were difficulties of access to a sufficient number of identical ponds or raceways, avoidance of interfering diseases and above all, obtaining the natural challenge at the right time.

The killed immersion vaccine has been extensively tested in the field in parallel with the laboratory trials. Not surprisingly, since its potency was doubtful in laboratory trials, field trials using the immersion delivery have been unsuccessful.

LIMITATIONS OF PRESENT VACCINES AND THE WAY AHEAD

Limitations of killed vaccines stem from the failure of immersion delivery. Injection of the vaccine into susceptible fingerlings is not commensurate with economic and management practices.

Limitations to live vaccines stem firstly from their residual (or potentially residual) virulence resulting in fears from fish farmers (to buy a product which may kill their fish) and from state authorities (live vaccines are legally prohibited almost everywhere). Secondly from an economic point of view, the above considerations do not create an encouraging background for vaccine manufacturers, more especially as the size of the potential market is rather small.

Scientific and technical facets

Killed vaccines

The problem with present killed immersion vaccines probably lies in the lack of effective uptake. Investigations on the uptake of the virus in close cooperation between gill physiology laboratories and fish virology laboratories could improve their present status.

Live vaccines

Mechanisms of virulence and host resistance

Currently we lack knowledge concerning the molecular basis of VHSV virulence. The present trends are to fill this gap in order to improve the means of selecting attenuated and immunogenic virus variants. Such data are also mandatory to obtain permission from the animal health authorities to allow the use of live vaccines for fish. Furthermore, an understanding of the virulence mechanisms could provide information on the factors hindering the onset of immunity. These factors are the same as those influencing the fish's susceptibility to disease and a better understanding of their effects can provide very useful data for control of VHS by further genetic selection of the host species.

Vestergaard-Jorgensen[21] pointed out that decreased persistence of the vaccine strain resulted in a decrease of vaccine potency. This suggests that to elicit protection, a vaccine strain must persist for as long as possible but also grow as slow as possible to avoid the onset of pathogenic or lethal infection. Difficulties of vaccination with attenuated VHSV strains lie in this balance which is also dependent on the fish's individual response.

Overcoming limitations

Producing an attenuated strain with the above characteristics is a worthwhile goal but the means of selecting such a strain depend upon methods emanating from an understanding of the mechanisms determining those characteristics. At present this understanding is lacking and if persistence and growth rate are intimately linked it may be impossible to obtain the ideal strain. However, since multiplication of vaccine strain F25 (21) immediately after entry into fish is significantly slower than the wild type 07.71,[23] it is possible that vaccine strains trigger non-specific defence mechanisms in a different way to wild strains. As hinted previously, interferon may play a role in eliminating the vaccinating virus. Until now, attenuation of VHS based upon increased induction of interferon synthesis could not be established[25] but our methods of interferon detection have recently been improved and investigations will be restarted.

Since the *in vivo* growth of attenuated virus strains seems to be hindered by non-specific immunity, probably interferon synthesis, an attenuated strain ideal for inducing specific immunity may be one which is poor in inducing interferon synthesis (unless its pathogenicity is thereby increased). *In vitro* methods to detect such variants could easily be developed.

Finally, stimulation of the specific immune response by most live vaccines is poorly understood and the evidence suggests that the protection observed may partially be due to non-specific mechanisms. Thus a further aim of future work should be to produce an attenuated strain which induces high levels of protecting antibody.

The duration of protection induced by the different anti-VHS live vaccines is shorter than the economic life of fish (which may also result from the effects of interferon), unless it is boosted by challenge. In this case resistance to VHS becomes as strong as that resulting from natural infection. Thus, in practice, fish should be transferred to infected areas soon after the vaccine delivery in order to maintain and enhance the immunity.

Field trial procedures

The absence of satisfactory results in field trials is a problem which could be solved if, instead of vaccination trials involving treated and untreated fish groups in the same trout farm facilities which are subject to various hazards as outlined above, one used the epidemiological method according to Michel *et al.*[9] This method consists in managing, on a large scale, the surveillance of pilot trout farms. In some, all fish would be vaccinated, and in others, none. The only participation expected from the fish farmer would be his agreement to allow health surveillance of the fish to assess the occurrence of VHS. Within 3 years such a programme applied to VHS infected farms would provide sufficient data to assess the economic efficacy of anti-VHS vaccination. The procedure would, however, involve obtaining permission of the state authorities implementing vaccine licensing. Indeed, since significant amounts of vaccine would be necessary, this will demand a temporary licence.

Modern approaches to vaccine development

In the light of modern molecular biology, the present vaccines look somewhat old fashioned and one can envisage the production of synthetic or genetically engineered vaccines. However, such vaccines will necessitate a high research investment and taking into account what is known with rabies virus, it is most unlikely that synthetic vaccines would be immunogenic without being conjugated to a carrier, for example to liposomes. Furthermore the difficulty of making the antigen or antigen-carrier complex effectively immunogenic by the immersion route will still persist and one will face the same problem with an expensive product as one already faces with the cheaper killed vaccine.

Legal facets of anti-VHS vaccination

From a regulatory view point, the use of live vaccines is prohibited in many countries, such as France. However, once a country or a watershed is heavily infected with VHS a control policy alone, using slaughter of stock and disinfection, will not ensure the eradication of the disease unless certain other measures are taken, namely – significant financial support and full willingness of fish farmers to cooperate. Furthermore, it assumes the watershed itself can be cleared of carrier feral or wild fish. These prerequisites are rather hard to fulfil.

A major criticism made from the regulatory viewpoint is that once a large-scale vaccination programme is begun it must be continued because of the continued source of the virus from carriers. It is obvious that, practically speaking, the goals of anti-VHS vaccines are to prevent disease but not necessarily infection. However, in heavily infected areas the real choice is between producing market size VHS carriers or losing a large proportion of the trout beforehand and the survivors still being carriers. It must be borne in mind that VHS variant strains which have been selected constitute a less serious threat for fish than the wild strains they originated from and that reluctance to use such strains as vaccines, based upon the risk of carriers and/or reversion, is not very realistic in countries or districts where VHS is widespread. However, even if legal limitations are overcome the problem then is to produce the vaccine commercially.

Economic facets of anti-VHS vaccination

From an economic view-point, the size of the potential market is a decisive factor in developing or manufacturing an anti-VHS vaccine. As a concrete example, in France the anti-VHS vaccine market is about 5×10^7 'doses' of live vaccine per year on the basis of 10^4 tonnes of trout (average weight 200 g) i.e. one third of the annual production, which represents the part of the yearly production reared in high-risk VHS areas. In Denmark, the market could be 2×10^7 doses (taking into account the success of the national disease control programme) and in Italy, slightly more than in France. On the basis of 0.2–0.3 UK pence per dose which is the cost range acceptable according to the French trout farmers, it does not make a great deal of money for a manufacturer with respect to the constraints that it represents! Beside the legal restrictions, the cost of production is also a deciding factor which in the past has made manufacturers somewhat cautious towards VHS vaccines even in heavily infected countries. In fact, according to our experience, the minimum concentration of the immersion vaccine is 1×10^5 pfu/ml with 1 litre of vaccine/1 kg fish, i.e. 1×10^{13} pfu for 5×10^7 2 g fish (representing a further yield of 10^4 tonnes of 200 g fish). Considering that in mass cell culture, the VHSV yield is at least 10^2 pfu/EPC cell, 1×10^{11} cells are needed for such a production which represents between 50 to 60 litres of vaccine! Thus, using improved virus production techniques VHS vaccine production is readily achievable both technically and economically.

CONCLUSIONS

Presently, there are no licensed anti-VHS vaccines although the potency of several preparations is well established. It is obvious that the residual virulence of the live vaccines makes them less than perfect but will perfect anti-VHS live vaccine exist one day? Indeed virulence depends as much on genetic factors in the pathogen as in the host and at this time one cannot predict what the susceptibility of the fish will be in the future. However, it is possible to obtain safe live vaccines at the expense of their potency.[25] Since the trout farming industry cannot afford to bear the charges of costly drugs, live vaccines are the cheapest which can be offered and they are ready for production. Synthetic or bioengineered vaccines would be perfect from the view-point of innocuousness but they do not exist and even if they did, they will have to prove their immunogenicity under a practical form of delivery.

Bearing in mind that certain changes in trout production and dealing could stem the spread of VHS better than anything else, it can be assumed that, at present, anti-VHS vaccination is an emergency control method which could be used in heavily infected districts and watersheds in which control policy alone cannot eliminate the disease. Of course, for the restocking of waterbodies and rivers, VHSV-free fish should be the only fish used. While vaccination could immediately contribute to the control of VHS, such a venture cannot be undertaken without the support and the firm request of the fish farmers affected by VHS because of the restrictive regulations.

For the future it remains evident that the market for an anti-VHS vaccine offers local and limited prospects. In the long term the best way of improving the resistance of fish to VHS and of stemming the spread of the disease may be genetic control associated with health control programmes. Taking this view into the future, the control methods for VHS should more closely resemble those used to combat plant diseases rather than animal diseases.

REFERENCES

1 Hill, B. J. (1987). Viral diseases of salmonid fish in Europe. *Proceedings of the International Symposium, Ichthyopathology in Aquaculture,* Dubrovnik, 21–24 October, 1986. (in press).

2 Olesen, N. J. and Vestergaard-Jorgensen, P. E. (1982). Can and do herons serve as vectors for Egtved virus? *Bull. Eur. Assoc. Fish Pathol.,* **2**, 48.

3 Vestergaard-Jorgensen, P. E. (1982). Egtved virus: occurrence of inapparent infections with virulent virus in free living rainbow trout. *J. Fish Dis.,* **5**, 251–255.

4 De Kinkelin, P. and Le Berre, M. (1977). Isolement d'un rhabdovirus pathogène de la Truite Fario (*Salmo trutta*). *Compte-Rendu de l'Académie des Sciences,* Paris, **284**, série D, 101–104.

5 Meier, W. and Vestergaard-Jorgensen, P. E. (1980). Isolation of VHS virus from pike fry (*Esox lucius*) with haemorrhagic symptoms. In: *Fish Diseases,* ed. W. Ahne, Proceedings in Life Sciences, Berlin: Springer Verlag, pp. 8–17.

6 Wizigmann, G., Baath, C. and Hoffmann, R. (1980). Isolierung des virus der viralen hämorrhagischen septikämie (VHS) aus Regenbogenforellen, Hecht und Aschenbrut. *Zbl. Vet. Med.,* B, **27,** 79–81.

7 Ahne, W. and Thomsen, I. (1985). Occurrence of VHS virus in wild white fish (*Coregonus* sp.). *Zbl. Vet. Med.*, B, **32**, 73–75.
8 Castric, J. and de Kinkelin, P. (1984). Experimental study of the susceptibility of two marine fish species, sea bass (*Dicentrarchus labrax*) and turbot (*Scophthalmus maximus*), to viral haemorrhagic septicaemia. *Aquaculture*, **41**, 203–212.
9 Michel, C., Tixier, G. and Mevel, M. (1984). Evaluation of the protective immunity and economic efficacy of vaccines for fish. In: *Symposium of Fish Vaccination*, ed. P. de Kinkelin and C. Michel. Paris: Office International des Epizooties, pp. 75–96.
10 Boutry, E. (1984). Influence de doses sublétales d'ammoniac sur l'infection de la truite arc-en-ciel par le virus d'Egtved. Diplôme d'Ingénieur d'Université, Université des Sciences et Techniques de Lille.
11 Chilmonczyk, S. and Oui, E. (1988). The effects of gamma irradiation on the lymphoid organs of rainbow trout and subsequent susceptibility to fish pathogens. *Veterinary Immunology and Immunopathology*, **18**, 173–180.
12 Dorson, M. and de Kinkelin P. (1974). Mortalité et production d'interféron circulant chez la truite arc-en-ciel après infection expérimentale avec le virus d'Egtved: influence de la température. *Annales de Recherches Vétérinaires*, **5** (3), 365–372.
13 De Kinkelin, P., Dorson, M. and Hattenberger-Baudouy, A-M. (1982). Interferon synthesis in trout and carp after viral infection. *Developmental and Comparative Immunology* **Suppl. 2**, 167–174.
14 De Kinkelin, P. and Le Berre, M. (1977). Démonstration de la protection de la Truite Arc-en-Ciel contre la SHV, par l'administration d'un virus inactivé. *Bulletin de l'Office International des Epizooties*, **87**, (5–6), 401–402.
15 Le Berre, M., de Kinkelin, P. and Metzger, A. (1977). Identification sérologique des rhabdovirus de salmonidés. *Bulletin de l'Office International des Epizooties*, **87** (5–6), 391–393.
16 Ahne, W., Vestergaard-Jorgensen, P. E., Olesen, N. J., Schäfer, W. and Steinhagen, P. (1986). Egtved virus: occurrence of strains not clearly identifiable by means of virus neutralization tests. *Journal of Applied Ichthyology*, **2**, 187–189.
17 De Kinkelin, P., Bernard, J. and Hattenberger-Baudouy, A-M. (1984). Immunization against viral diseases occurring in cold water. In: *Symposium on Fish Vaccination*, ed. P. de Kinkelin and C. Michel. Paris: Office International des Epizooties, pp. 167–197.
18 Ghittino, P., Schwedler, H. and de Kinkelin, P. (1984). The principal infectious diseases of fish and their general control measures. In: *Symposium on Fish Vaccination*, ed P. de Kinkelin and C. Michel. Paris: Office International des Epizooties, pp. 5–38.
19 Dorson, M. and Chevassus, B. (1985). Etude de la réceptivité d'hybrides triploïdes truite arc-en-ciel × saumon coho à la nécrose pancréatique infectieuse et à la septicémie hémorragique virale. *Bulletin Français de Pêche et Pisciculture*, **296**, 29–34.
20 Vestergaard-Jorgensen, P. E. (1976). Partial resistance of rainbow trout (*Salmo gairdneri*) to viral haemorrhagic septicaemia (VHS) following exposure to non-virulent Egtved virus. *Nord. Veterinary Medicine*, **28**, 570–571.
21 Vestergaard-Jorgensen, P. E. (1982). Egtved virus: temperature-dependent immune response of trout to infection with low-virulence virus. *J. Fish Dis.*, **5**, 47–56.

22 De Kinkelin, P., Bearzotti-Le Berre, M. and Bernard, J. (1980). Viral haemorrhagic septicaemia of rainbow trout: selection of a thermoresistant virus variant and comparison of polypeptide synthesis with the wild type virus strain. *Journal of Virology,* **36**, 652–658.
23 De Kinkelin, P. and Bearzotti-Le Berre, M. (1981). Immunization of rainbow trout against viral haemorrhagic septicaemia (VHS) with a thermoresistant variant of the virus. *Developments in Biological Standardization,* **49**, 431–439.
24 Bernard, J., de Kinkelin, P. and Bearzotti, M. (1983). Viral haemorrhagic septicaemia of trout: relation between the G polypeptide, antibody production and protection of the fish following infection with the F25 attenuated variant strain. *Infection and Immunity,* **39**, 7–14.
25 Bernard, J., Bearzotti-Le Berre, M. and de Kinkelin, P. (1985). Viral haemorrhagic septicaemia in rainbow trout: attempt to relate interferon production, antibody synthesis and structure of the virus with the mechanism of virulence. *Annales de Virologie (Institut Pasteur),* **136 E**, (1), 13–26.
26 Dorson, M. and Torchy, C. (1979). Complement dependent neutralization of Egtved virus by trout antibodies. *J. Fish Dis.* **2**, 345–347.
27 Fijan, N., Sulimanovic, D., Bearzotti, M., Muzinic, D., Zwillenberg, L. O., Chilmonczyk, S., Vautherot, J. F. and de Kinkelin, P. (1983). Some properties of the *Epithelioma papulosum cyprini* (EPC) cell line from Carp, *Cyprinus carpio. Annales de Virologie* (*Institut Pasteur*), **134 E**, 207–220.
28 Wolf, K. and Quimby, M. C. (1962). Established eurythermic line of fish cell *in vitro. Science,* **135**, (3508), 1065–1066.
29 Vestergaard-Jorgensen, P. E. (1972). Egtved virus: antigenic variation in 76 virus isolates examined in neutralization tests and by means of the fluorescent antibody technique. In: *Symposia of the Zoological Society of London* **30**, 333–339, Academic Press.
30 Ghittino, P. (1968). Grave enzootia di setticemia emorragica in trotte fario di allavamento (*Salmo trutta*). *Rivista italiana di Piscicoltura e Ittiopatologia,* **1**, 17–19.

15

Vaccination against Infectious Haematopoietic Necrosis*

J. C. Leong, J. L. Fryer and J. R. Winton

THE DISEASE PROBLEM

Economic significance – an overview

Infectious haematopoietic necrosis is a disease of certain salmonid fish native to the North Pacific Rim. The disease was first reported in 1953 as the cause of mortality among sockeye salmon (*Oncorhynchus nerka*) in the state of Washington.[1] Since that time, infectious haematopoietic necrosis virus (IHNV) has been recovered from susceptible salmonid species from Alaska to Northern California[2,3] and numerous epizootics have occurred among hatchery reared fish within this range. In 1972, the virus was recovered from sockeye salmon in Hokkaido, Japan.[4] IHNV has also been found in rainbow trout (*Salmo gairdneri*) and chum salmon (*Oncorhynchus keta*) on Honshu where it has spread widely. More recently the virus has been recovered from stocks of rainbow trout in Taiwan.[5] The authors are unaware of any reports of isolations of IHNV from the Soviet Far East or from Korea. While the most severe IHNV epizootics have been observed in hatcheries, natural outbreaks of IHNV have also been reported. The virus was isolated from young sockeye salmon at Chilko Lake, British Columbia, where 10–20 per cent of the salmon fry in the lake died of IHN.[6]

Since its discovery, the incidence of IHNV has increased dramatically, and the only available control measure for the disease has been destruction of the infected fish population. Because of the economic impact of IHNV on stocks of salmon and trout, development of vaccines and appropriate delivery systems have become important areas for research.

Populations at risk

IHNV produces a severe disease among fry and juvenile fish. Approximately 4 days after exposure to the virus, the infected fish become darker and show

*Oregon Agricultural Experiment Station Technical Paper No. 8249.

FISH VACCINATION
ISBN 0-12-237485-1

haemorrhages at the bases of fins. In the advanced stages of the disease, petechial haemorrhages appear in the musculature and in the mesenteries of the internal organs. Pathological examination of the infected fish indicates that the kidneys and spleen are the major target organs.

An epizootic of IHN is usually characterized by an abrupt onset and high mortality among the infected population. In fish up to 2 months of age, the mortality often exceeds 90 per cent. As fish become older susceptibility to IHNV decreases. Mortality is unusual in fish 2 years of age or older.

While virus may replicate *in vitro* between 5–20°C, most epizootics occur in the range of 10–14°C. A temperature sensitive strain has been recovered from fish at one location.[7] IHN has been controlled in fish at this site by raising the water temperature to 18°C; however, the number of carriers within the population has remained high.[8]

Transmission of IHNV

The route of transmission of IHNV in nature is uncertain. However, experimental studies have shown that the virus can be transmitted both horizontally and vertically. Horizontal water-borne transmission of IHNV has been demonstrated by a number of investigators.[9–11] In these studies, susceptible fish held in the effluent water from fish infected with IHNV were subsequently infected with IHNV. The initial sites of infection were the gills. The infection spread to visceral target organs such as the anterior kidneys and spleen. Infection by the oral route has been demonstrated by feeding the carcasses of infected fry to susceptible fish.[9,12]

The strongest evidence for vertical transmission comes from the association made between the appearance of the disease and the shipment of eggs from infected adults into geographical areas where the virus was not known to occur.[13–15] Mulcahy and Pascho[16] recently demonstrated IHNV transmission in naturally infected eggs and fry which were raised in virus-free water. However, Groberg[17] reported that when fertilized eggs were treated with iodophore and incubated to hatching in virus-free water, no virus was isolated nor signs of disease observed.

Carriers of IHNV

The existence of an IHNV carrier state is a contested issue among many scientists. In 1975, Amend[9] reported that in rainbow trout surviving an IHNV epizootic, the virus entered a latent state and was not recoverable until these fish reached sexual maturity. The virus was then found in milt or ovarian fluids. Other investigators have also attempted to detect a carrier state by other methods with little success (Groberg, pers. comm.; Mulcahy, pers. comm.).[18] The ability to recover IHNV from fish surviving an epizootic only at the time of sexual maturity, raises the question as to whether the sexually mature fish expressed a latent virus or if viral reinfection occurred from some other reservoir.

If fish are reinfected with IHNV at spawning time, other reservoir(s) must be identified. The virus has been found to be extremely stable in sediments which may serve as sources of contamination (Mulcahy, pers. comm.). Other possible hosts, including resident wild fish and fish parasites, have been examined. Since the nature of the carrier state and the potential reservoirs of IHNV are essential elements in determining how and when a vaccine should be applied, they are important areas for research.

Pathogen heterogeneity

A serological comparison of rainbow trout, sockeye and chinook salmon isolates of IHNV using polyvalent rabbit antisera to whole virus indicated that while all isolates were similar, the chinook strain appeared different.[19] More recently, monoclonal antibodies (Mab) with neutralizing activity were used to distinguish among different virus strains.[20] At least three serological strains of IHNV were identified in this way.

Another method used for identifying strain differences among virus isolates was based on the differences in the migration patterns of the viral proteins on sodium dodecyl sulphate polyacrylamide gels (SDS–PAGE).[21] The virion protein patterns of 71 isolates of IHNV were analysed by SDS–PAGE of ^{35}S-methionine labelled intracellular viral proteins. With the molecular weights of the viral glycoproteins and nucleoproteins as a basis, five groups were identified. This method provided a reproducible and rapid way to identify virus isolates. In addition, the method permitted investigators to determine that a particular virus type may be endemic to an area and that different salmonid species within the area carry the same virus type.

Possible chemotherapeutic measures

At the present time there is no licensed chemotherapeutic drug for the control of IHNV infections in fish. However, research has been conducted in Japan to identify antiviral drugs for treatment of IHNV-infected fish. In a survey of 24 possible antiviral drugs, Hasobe and Saneyoshi[22] identified four compounds with *in vitro* anti-IHNV activity. These compounds were tested *in vivo* in IHNV infected steelhead fry. Only two compounds, 6-thioinosine and 5-hydroxyuridine, produced a modest reduction in mortality.

IHNV VACCINE

Types of vaccines

There are three types of vaccines that have been considered for IHNV control: attenuated, killed, and subunit viral vaccines produced by recombinant DNA technology. All three have had some success in inducing protective immunity in fish.

Among the selected criteria that an IHNV vaccine must meet, safety is the most important. A vaccine used in a system with an untreated outflow must be completely harmless for all potentially affected species in the watershed. Thus, an attenuated vaccine must be avirulent for resident fish and aquatic organisms. Also, the reversion frequency to virulence for a live vaccine must be carefully determined.

An ideal IHNV vaccine must also demonstrate effectiveness. Because fish are susceptible to lethal infection as early as newly-hatched fry and as late as yearlings, the vaccine should provide protective immunity during this period. It is a serious concern that the fish are most susceptible at a stage in their development when the immune system may be incapable of responding to antigenic stimulation. Leong[23] has shown that rainbow trout as small as 0.4 g are capable of mounting a protective immune response to IHNV. However, the duration of immunity to IHNV induced by the various types of vaccines has not been adequately ascertained.

Other important considerations in the development of an IHNV vaccine are those of cost and ease of administration. An ideal IHNV vaccine should be inexpensive to prepare and have a stable shelf-life. The vaccine must be easy to administer to large numbers of fish. It must provide protective immunity against all strains of IHNV and in all susceptible host species.

Vaccine trials

Laboratory trials have been conducted with the attenuated, killed and subunit IHNV vaccines. These studies have provided encouraging results which suggest that vaccination of salmonid fishes against the virus is possible. No field trial data have been reported. However, large-scale tests of vaccines conducted by private and public agencies are known to have yielded mixed results. In the laboratory, an attenuated IHN vaccine derived from a rainbow trout isolate was developed at Oregon State University. This vaccine provided protection of kokanee salmon against water-borne or intraperitoneal challenge with wild type virus (Table 15.1). Protection was found to last for at least 110 days.[24] Additional studies[25] showed chinook salmon could also be protected with the same vaccine. When rainbow trout were vaccinated, however, significant residual virulence was detected.[26]

Table 15.1 *Immunization of sockeye salmon with an attenuated vaccine against IHNV*[24]

Treatment	Challenge	Per cent mortality
Vaccinated	Injected	5
	Water-borne	5
Nonimmunized controls	Injected	100
	Water-borne	90

Sockeye salmon (0.9 g) were immunized by exposure to 6200 pfu/ml of water containing the attenuated vaccine for 48 h. After 25 days, fish were challenged by intraperitoneal injection of 40 LD_{50} or by exposure in water to 4×10^3 pfu/ml wild type IHN virus.

Further attenuation and testing of the virus was suspended due to the lack of commercial interest.

Two groups have reported success with vaccination of rainbow trout using killed preparations of IHNV. Amend[27] used β-propiolactone to inactivate IHNV and immunized trout fry by intraperitoneal injection. He found the immunized fish were resistant to a lethal challenge dose of IHNV (Table 15.2). More recently, Nishimura *et al.*[28] tested several methods of formalin inactivation of IHNV to determine the optimum inactivation schedule. Several of the formalin-killed vaccines were used to protect juvenile trout against lethal challenge. The vaccine was most effective when delivered by injection; however, hyperosmotic immersion was capable of stimulating limited immunity (Table 15.3).

Table 15.2 *Immunization of rainbow trout with a β-propiolactone-killed vaccine against IHNV*[27]

Treatment	Challenge	Per cent mortality
Vaccinated	Injected	4
Nonimmunized controls	Injected	70

Rainbow trout were injected intraperitoneally with 0.05 ml of vaccine made by adding β-propiolactone (1 : 6000) to a suspension containing $10^{7.3}$ $TCID_{50}$/ml IHNV. After 32 days, fish were challenged by subcutaneous injection of $10^{6.2}$ $TCID_{50}$ of wild type virus. The size of the fish at the time of immunization was not given.

A subunit vaccine derived from a cloned sequence of the glycoprotein gene[29] has been tested in the laboratory.[30] In trials using injected and water-borne delivery, protection against a water-borne challenge was obtained when vaccinates were compared with controls (Tables 15.4 and 15.5). This preparation has been shown to be capable of stimulating immunity in at least two species of host fish (rainbow trout and chinook salmon) and is effective against heterologous isolates (Table 15.5).

Table 15.3 *Immunization of rainbow trout with a formalin-killed vaccine against IHNV*[28]

Treatment	Challenge	Per cent mortality
Inoculated	Injected	5
Noninoculated controls	Injected	75
Immersed	Water-borne	50
Nonimmersed controls	Water-borne	80

Rainbow trout (2.5 g) were injected intraperitoneally with 0.05 ml of vaccine made from viral suspension containing 10×10^{10} $TCID_{50}$/ml inactivated with formalin (0.1 to 0.8 per cent) and challenged after 28 days by injection of $10 \times 10^{5.1}$ $TCID_{50}$/ml. Immersion vaccination of rainbow trout fry (0.9 g) was in a hyperosmotic solution containing 5 per cent inactivated vaccine in 5 per cent NaCl for 2 min. These fry were given a second immunization by immersion after 14 days. Two weeks after the booster immunization, these fish were challenged by immersion in $10 \times 10^{4.1}$ $TCID_{50}$/ml of wild type virus.

Table 15.4 *Immunization of rainbow trout with a subunit vaccine against IHNV*[30]

Treatment	Challenge	Per cent mortality
Inoculated[a]	Water-borne	9
Immersed[b]	Water-borne	36
Nonimmunized controls	Water-borne	60

[a]Rainbow trout (0.4 g) were injected with 0.01 ml of a bacterial lysate (3 mg/ml) containing 8 per cent expressed G protein.
[b]Rainbow trout (0.4 g) were immersed in the same bacterial lysate (50 fish per 10 ml of solution) for 3 min. After 1 month, fish were challenged by water-borne exposure to various concentrations of virus. Data shown are for a challenge with 3.2×10^3 $TCID_{50}$/ml.

Table 15.5 *Immunization of rainbow trout and chinook salmon with subunit vaccine against IHNV. Homologous* vs *heterologous challenge*[30]

Treatment	Challenge	Per cent mortality
Immersed	Round Butte IHNV[a]	19
Control	Round Butte IHNV[a]	92
Immersed	Elk River IHNV[b]	0
Control	Elk River IHNV[b]	64
Immersed	Dworshak IHNV[c]	11
Control	Dworshak IHNV[c]	85

[a]Rainbow trout (0.4 g) were immersed in a bacterial lysate (3 mg/ml) containing 8 per cent expressed G protein for 3 min. After 1 month, the fish were challenged by water-borne exposure to various concentrations of virus. Data shown are for a challenge of 7.2×10^5 $TCID_{50}$/ml.
[b]Chinook salmon (0.4 g) were immunized by immersion as above. The data shown here are for a challenge of 1.93×10^6 $TCID_{50}$/ml.
[c]Rainbow trout (0.4 g) were immunized by immersion as above. The data shown here are for a challenge of 2.85×10^3 $TCID_{50}$/ml.

No field trials with IHNV vaccines have been reported in the literature, but a test was conducted in California, USA using an attenuated strain of IHNV. The vaccine in this trial did not give adequate protection. Attenuated vaccines have also been tested in Idaho, USA with varying success.

Effectiveness

As side-by-side trials have not been reported, there is insufficient information to determine the comparative efficacy of the various types of IHN vaccines under development. However, differences in the effectiveness of each vaccine have been found to be dependent upon the method of delivery. When administered by immersion, modified live IHN vaccines have been effective and generally provide better immunity than an inactivated vaccine. This result may be expected because the replication of the attenuated virus stimulates additional forms of interference and immunity. This modified live vaccine has provided protection against both injected and water-borne challenge with wild type virus.[24]

When delivered by injection, killed vaccines have shown protection of fish against both types of challenge.[28] The subunit vaccine delivered by injection or immersion confers protection against a water-borne challenge.[30] The duration of the immunity and the effect of a booster dose on immunity has not been well established for any vaccine.

Limitations

The principal limitations for IHN vaccines involve safety, efficacy and cost. Attenuated vaccines must be avirulent and free of any potential for reversion via back mutation or recombination to a more virulent form. Reversion has been a concern for other live modified vaccines against rhabdoviruses (e.g. rabies). Both killed and subunit vaccines against IHNV administered by injection have demonstrated efficacy. However, this delivery method may prove too expensive for the vaccination of millions of small fish. Efficacy has also been demonstrated for the subunit vaccine when administered by immersion. The subunit vaccine will be less expensive to produce than a killed whole virus vaccine.

One potential limitation of all IHN vaccines may be the degree of cross-protection afforded fish against the biochemically and serologically disparate strains of IHNV that occur in different species of salmonids or in various geographic areas. If a cross-protection limitation is shown to be significant, the development of several IHN vaccines for use in each application (polyvalent preparations) will be required. Experimental studies, however, indicate that the recombinant subunit vaccine will confer protection against challenge by at least two different isolates of IHNV (Table 15.5). More recently, we have shown that purified IHNV (Round Butte strain) glycoprotein, prepared from whole virus, will confer immunity to the five biochemically characterized IHNV strains (Engelking and Leong, unpublished results).

A final limitation may be the size of the commercial market. In order to defray the research and licensing costs necessary for the production of IHN vaccines, a sufficient market must be developed. The market will be determined by the efficacy and cost of the vaccine. Since in most applications millions of small and relatively inexpensive fish need protection, the cost per dose must be low. Also, the market will be restricted to portions of North America and Japan where IHNV is found.

Overcoming limitations

The limitations to the development of a safe, effective and inexpensive vaccine to protect salmonid fish against IHNV will be difficult to overcome. Important areas for research are the types of immunity stimulated by different viral vaccines, the use of adjuvants in enhancing immunity, and the role of booster immunizations. Attenuated vaccines will have to be proven stable and of low virulence, especially for the relatively small fish that must be protected against disease. The number of serotypes of IHNV must be determined. For inactivated and subunit vaccines, effective delivery systems to replace time-consuming injections should be developed.

For any of the IHNV vaccines under development or produced in the future, adequate markets must be assured to provide incentive for research and required licensing. Vaccines must have a low dosage cost and be easy to administer. Meeting these needs will require research and a better understanding of fish immunology.

Commercial prospects

In spite of the limitations of the vaccines currently being tested, the commercial development of an effective IHN vaccine appears promising. The regulatory process for the licensing of fish vaccines is improving, and salmonid aquaculture within the enzootic area is increasing. The spread of IHNV and its increased virulence for larger, more valuable trout will enlarge the market for any vaccine available.

The easiest vaccine to develop, test, and license may be a killed whole virus preparation, but the cost per dose would certainly be high. A subunit vaccine would be less expensive, and if efficiently delivered, relatively easy to license. The vaccine industry has expressed concern about licensing requirements for attenuated vaccines and their acceptance by fish pathologists, who rely on viral detection methods for certification of fish stocks, remains uncertain. Attenuated preparations also pose a problem of maintaining control for vaccine producers. Without a specific viral market and proper patent protection, it would be easy for anyone to purchase and produce the vaccine strain.

ACKNOWLEDGMENTS

This research was supported by US Dept of Energy – Bonneville Power Administration Contracts No. DE-AI79-83BP11987 and No. DE-AI79-84BP16479. Oregon Agricultural Experiment Station Technical Paper No. 8249.

REFERENCES

1 Rucker, R. R., Whipple, W. J., Parvin, J. R. and Evans, C. A. (1953). A contagious disease of salmon possibly of virus origin. *US Dept Interior Fish Wildl. Serv. Fish. Bull.*, **76** (54), 35–46.
2 Pilcher, K. S. and Fryer, J. L. (1978). The viral diseases of fish: a review through 1978. Part II: Diseases in which a viral etiology is suspected but unproven. *CRC Crit. Rev. in Micro.*, **7** (4), 287–364.
3 Grischkowsky, R. S. and Amend, D. F. (1976). Infectious hematopoietic necrosis virus: prevalence in certain Alaskan sockeye salmon (*Onchorhynchus nerka*). *J. Fish Res. Board Can.*, **33** (1), 186–188.
4 Kimura, T. and Awakura, T. (1977). Studies on viral diseases of Japanese fishes. VI. Infectious hematopoietic necrosis (IHN) of salmonids in the mainland of Japan. *J. Tokyo Univ. Fish.*, **63** (2).
5 Chen, S. N., Chi, S. C. Shih, H. H. and Kou, G. H. (1983). The occurrence of infectious hematopoietic necrosis virus (IHNV) in cultured rainbow trout (*Salmo gairdneri*) in Taiwan. *Proceedings of Republic of China–Japan Cooperative Science Seminar on Fish Diseases.* 15–21 November 1982. Tungkang Marine Laboratory, Taiwan, ROC, pp. 56–58.
6 Williams, I. V. and Amend, D. F. (1976). A natural epizootic of infectious hematopoietic necrosis in fry of sockeye salmon (*Oncorhynchus nerka*) at Chilko Lake, British Columbia. *J. Fish Res. Board Can.*, **33**, 1564.
7 Mulcahy, D., Pascho, R. and Jenes, C. K. (1984). Comparison of *in vitro* growth characteristics of ten isolates of infectious hematopoietic necrosis virus. *J. G. Virol.*, **65** (12), 2199–2207.
8 Amend, D. F. (1970). Control of infectious hematopoietic necrosis virus disease by elevating the water temperature. *J. Fish Res. Board Can.*, **27** (2), 265–270.
9 Amend, D. F. (1975). Detection and transmission of infectious hematopoietic necrosis virus in rainbow trout. *J. Wildl. Dis.*, **11**, 471–478.
10 Wingfield, W. H. and Chan, L. D. (1970). Studies on the Sacramento River Chinook disease and its causative agent. In: *Symposium on Diseases of Fishes and Shellfishes*, ed. S. Snieszko, *Am. Fish. Soc. Spec. Publ.*, **5**, 307–318.

11 Mulcahy, D., Pascho, R. J. and Jenes, C. K. (1983). Titre distribution patterns of IHNV in ovarian fluids of hatchery and feral salmon populations. *J. Fish Dis.*, **6**, 183–188.
12 Watson, S. W., Guenther, R. W. and Rucker, R. R. (1954). A virus disease of sockeye salmon: interim report. *US Dept Interior Fish and Wildl. Serv. Special Scientific Report, Fisheries* **138**.
13 Holway, J. E. and Smith, C. E. (1973). Infectious hematopoietic necrosis of rainbow trout in Montana: a case report. *J. Wildl. Dis.* **9**, 287–290.
14 Plumb, J. A. (1972). *Some Biological Aspects of Channel Catfish Virus Disease.* Dept of Fisheries and Allied Aquaculture, Agricultural Experiment Station, Auburn University, Alabama, p. 1.
15 Wolf, K., Quimby, M. C., Pettijohn, L. L. and Landolt, M. L. (1973). Fish Viruses: isolation and identification of infectious hematopoietic necrosis in Eastern North America. *J. Fish. Res. Board Can.*, **30** (1), 1625–1627.
16 Mulcahy, D. and Pascho, R. J. (1985). Vertical transmission of infectious hematopoietic necrosis virus in sockeye salmon (*Oncorhynchus nerka*) (Walbaum): isolation of virus from dead eggs and fry. *J. Fish Dis.*, **8**, 393–396.
17 Groberg, W. J. (1987). Production trials of rearing progeny from two species of adult salmonids carrying infectious hematopoietic necrosis virus. *Abstract from the 24th Annual Symposium and Meeting,* Oregon Chapter, American Fisheries Society, Welches, Oregon. 28–30 Jan. 1987.
18 LeVander, L. J., Hopper, K. and Amos, K. H. (1985). IHNV transmission – a study of IHNV carrier state in sockeye salmon. *Abstract of 1985 Northwest Fish Culture Conference.*
19 McCain, B. B., Fryer, J. L. and Pilcher, K. S. (1971). Antigenic relationships in a group of three viruses of salmonid fish by cross neutralization. *Proc. Soc. Exp. Biol. and Med.,* **137** (3), 1042–1046.
20 Arakawa, C. K., Lannan, C. N. and Winton, J. R. (1986). Monoclonal antibodies recognize strains of IHNV. Fish Health Section Newsletter, *Am. Fish. Soc. Newsletter,* **14** (3), 1.
21 Hsu, Y. L., Engelking, H. M. and Leong, J. C. (1986). Occurrence of different types of infectious hematopoietic necrosis virus in fish. *Applied and Environmental Microbiol.,* **52** (6), 1353–1361.
22 Hasobe, M., and Saneyoshi, M. (1985). On the approach to the viral chemotherapy against infectious hematopoietic necrosis virus (IHNV) *in vitro* and *in vivo* on salmonid fishes. *Fish Pathol.,* **20** (2/3), 343–351.
23 Leong, J. C. (1985). Evaluation of a subunit vaccine to infectious hematopoietic necrosis (IHN) virus. *1984 Annual Report to Bonneville Power Administration,* Division of Fish and Wildlife, US Dept of Energy. July 1985.
24 Fryer, J. L., Rohovec, J. S., Tebbit, G. L., McMichael J. S. and Pilcher, K. S. (1976). Vaccination for the control of infectious diseases in Pacific salmon. *Fish Pathol.* **10** (2), 155–164.
25 Tebbit, G. L. (1976). Viruses infecting salmonid fishes from Oregon, A. The occurrence and distribution of infectious pancreatic necrosis virus, B. The development of an attenuated strain of infectious hematopoietic necrosis virus (IHNV) for the immunization of salmonids. Ph.D. Thesis, Oregon State University, Corvallis, Oregon, 81 pp.
26 Rohovec, J. S., Winton, J. R. and Fryer, J. L. (1981). Bacterins and vaccines for the control of infectious diseases in fish. In: *Proceedings of Republic of China–United States Cooperative Science Seminar on Fish Diseases,* NSC Symposium Series No. 3, National Science Council, Republic of China.

27 Amend, D. F. (1976). Prevention and control of viral diseases of salmonids. *J. Fish Res. Board Can.,* **33** (4), 1059–1066.
28 Nishimura, T., Sasaki, H., Ushiyama, M., Inoue, K., Suzuki, Y., Ikeya, F., Tanaka, M., Suzuki, H., Kohara, M., Arai, M., Shima, N. and Sano, T. (1985). A trial of vaccination against rainbow trout fry with formalin killed IHN virus. *Fish Pathol.* **20**, 435–443.
29 Koener, J. F., Passavant, C. W., Kurath, G. and Leong, J. C. (1987). Nucleotide sequence of a cDNA clone carrying the glycoprotein gene of infectious hematopoietic necrosis virus, a fish rhabdovirus. *J. Virology,* **61** (5), 1342–1349.
30 Gilmore, R. D., Engelking, H. M., Manning D. S. and Leong, J. C. (1988). Expression in *Escherichia coli* of an epitope of the glycoprotein of infectious hematopoietic necrosis virus protects against viral challenge. *Bio/Technology,* **6**, 295–300.

Note added in proof.

Since submission of this article, the following reports have been published.

Engelking, H. M. (1988). Properties of the glycoprotein of infectious hematopoietic necrosis virus. Doctoral Thesis, Oregon State Univ., Corvallis, USA.

Roberti, K. A. (1987). Variants of infectious hematopoietic necrosis virus selected with glycoprotein-specific monoclonal antibodies. Master's Thesis, Oregon State Univ., Corvallis, USA.

Winton, J. R., Arakawa, J. R., Lannan O. N. and Fryer, J. L. (1988). Neutralizing monoclonal antibodies recognize antigenic variants among isolates of infectious hematopoietic necrosis virus. *Diseases of Aquatic Organisms,* **4** (in press).

16

Vaccination against Spring Viraemia of Carp

N. Fijan

THE DISEASE PROBLEM

Spring viraemia of carp (SVC) is an acute contagious disease caused by *Rhabdovirus carpio* (or SVC virus) and was first described by Fijan *et al.*[1] Carp is the principal species affected and all age groups are susceptible. As the name implies, the disease typically occurs in spring, during rising water temperatures. The disease is often complicated by the simultaneous occurrence of bacterial and parasitic diseases as well as secondary infections.

Clinical signs include congregation of infected fish near pond banks or bottom, sluggishness, uncoordinated swimming, dark coloration, pale gills, petechial haemorrhages in the skin, eyes, gill and bases of fins, exophthalmia, extended belly and trailing pseudofaecal casts from protruding and inflamed anus. Internal examination sometimes reveals only an oedematous enlargement of organs and inflamed intestine. A varying degree of haemorrhaging is present in swimbladder, internal organs and muscles and the body cavity may contain fluid. Diagnosis is based upon culture and identification of SVC virus in the laboratory.

Geographic range

SVC has been diagnosed in Yugoslavia, France, Czechoslovakia, FR Germany, Austria, Hungary, USSR and Great Britain. Most probably, the disease is present in all other European countries having carp culture and possibly also in other continents where carp are reared under climatic conditions similar to those in Europe.

The virus is widespread in Europe: serological surveys demonstrated the presence of neutralizing antibodies on 95 per cent of carp farms in Austria,[2] 86 per cent of farms in Bavaria[3] and on 28 per cent of facilities in France.[4]

Endemicity of SVC was not studied, but the above percentages of serologically positive farms as well as the persistence of positivity over several years clearly indicate its existence.

FISH VACCINATION
ISBN 0-12-237485-1

Economic significance

The occurrence and severity of SVC outbreaks varies from year to year, from one farm to another and from pond to pond. Factors involved in such irregular epizootiological patterns are not well understood. Populations of fingerlings without neutralizing antibodies in autumn seem to be at a higher risk of serious disease outbreaks in spring. Mortality rates in outbreaks vary usually around 30–40 per cent, but sometimes exceed 70 per cent. The economic impact of SVC is therefore rather serious by elevating the price, through unavailability of fingerlings for restocking and by considerably affecting the annual yield.

Populations at risk

Carp is the common and principal victim of SVC. Natural outbreaks of the disease have been reported in crucian carp (*Carassius carassius*), grass carp (*Ctenopharyngodon idella*), bighead (*Hypophthalmichthys molitrix*), silver carp (*Aristichthys nobilis*), the fry of pike (*Esox lucius*) and of sheatfish[18] (*Silurus glanis*). Experimental infections have been induced in guppy (*Lebistes reticulatus*) and pumpkinseed (*Lepomis gibbosus*).

Although all age groups of carp are susceptible to SVC, the seasonality factor results in the principal victims being 9–12 and 21–24 months of age.

Season and temperature profoundly influence the host–virus interaction. The physiological status of carp after overwintering is thought to be a contributing factor to outbreaks of the disease in spring at the same temperatures as those in autumn, when severe losses never occur. Mortalities due to SVC occur from November–July, with a peak in April–June. In other words, disease outbreaks are rare below 10°C, most common between 11°C and 17°C and again rare between 17°C and 20°C. Mortalities cease at about 22°C, though some reports put this temperature limit as low as 17°C. Sheatfish fry can be affected at temperatures above 22°C.[19]

The influence of management practices and of water quality on the disease have not been studied sufficiently. Holding of fish in wintering ponds (high stocking density, usually without feeding) during the initial rise of temperature in spring and all the factors evoking stress favour SVC outbreaks and high mortalities. Typical SVC outbreaks occur about 10–20 days after transfer of fish to grower-ponds in spring. Change of management procedure to transfer of fish in autumn (no handling stress before the critical temperature rise) did not alleviate the risk of disease in the following spring.

The annual differences of pond water temperature in Europe can approach 30°C. Data from the literature on the carp–SVC virus interactions in this temperature range are schematically presented in Fig. 16.1.

These data suggest explanations about the epidemiological pattern of SVC and about the lack of effectiveness of changing management practice. The clinical disease occurs in spring at the lower temperature limit for interferon and antibody production by carp. Presumably the rapid production of such factors during summer and the persistence of antibodies through the autumn protect the fish at this time of the year. The length of overwintering at low temperatures may be conducive to catabolism of Ig so that by spring no protective antibody is present whether or not fish transfer occurs in spring or in autumn.

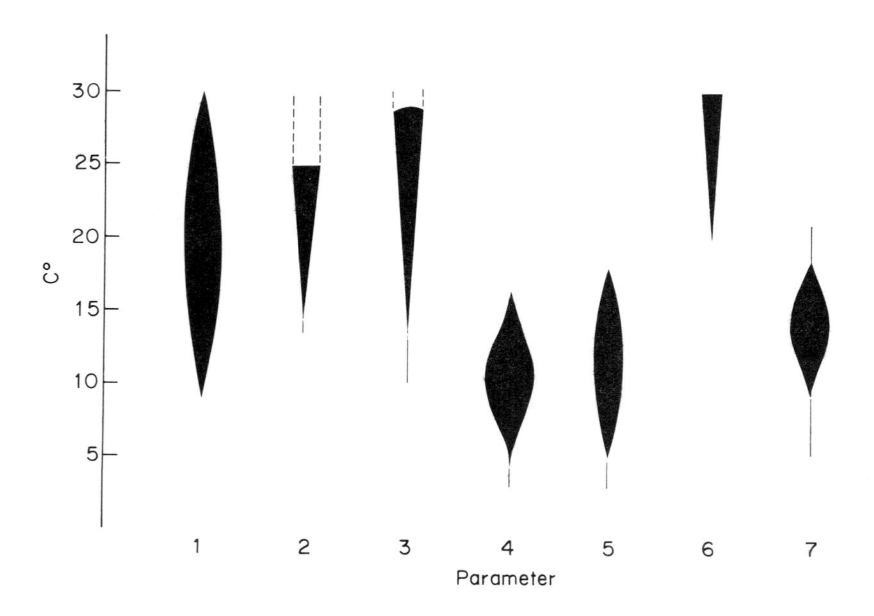

Fig. 16.1 Influence of temperature on carp–*Rhabdovirus carpio* interactions (from Fijan[17]). 1 = Replication of virus *in vitro*; 2 = Induction of interferon by virus; 3 = Induction of neutralizing antibodies; 4 = Long lasting viraemia; 5 = Detectable shedding of virus; 6 = Undetectable persistence and shedding of virus (not proved experimentally); 7 = Clinical disease and mortality.

Transmission

Reservoirs of infection include carriers, diseased fish and carcasses. Increases in the incidence of antibody-positive carp during late summer indicate the existence of healthy carriers and chronic post-epizootic carriers. Wild fish entering ponds with water are strongly suspected to be carriers of virus. Proved vectors are the carp louse and leeches,[5] water and mud. The virus remains infective in water at 10°C for about 35 days, in mud at 4°C for 42 days and in dried mud at constant temperatures between 4° and 20°C for more than 4 weeks.[6] The role of birds in transmission has not been investigated and the range of fish species acting as carriers or vectors as well as the factors influencing virus shedding are not well known. Thus, the lateral transfer by water, mud, vectors and equipment seems to be the principal mode of virus transmission. There is little evidence for frequent vertical transmission. Fijan *et al.* (unpublished) did not find SVC virus in sexual products of 50 females and 36 males from farms affected by the disease. Similar findings were reported by Ahne.[5] Békesi and Csontos[7] detected only three positive ovarian fluids among 702 samples from both sexes. These findings, as well as the lack of natural SVC outbreaks among carp larvae and fry, indicate vertical transmission to be of minor importance. Factors affecting the length of carrier status and virus shedding in carp broodfish, other age classes and in other species have not been studied.

Pathogen heterogeneity

Ahne[8] described serological and pathogenic characteristics of one isolate (V 77) which could be regarded as a SVC subtype and another isolate (V 76) which is more closely related to pike fry rhabdovirus (PFR) than to SVC virus. These and other findings have led to the speculation that both the PFR and the above subtypes may represent variants of SVC virus. Somewhat serologically different virus isolates have occasionally been encountered by others (Fijan, unpublished; Kölbl, unpublished) but they have not yet been studied.

Limitations of control by means other than vaccination

SVC can be eradicated by slaughter of stock and disinfecton of premises on small farms and in indoor facilities. Eradication has little chance of success on most medium and large European farms[19] because it is presently impossible to completely prevent movements of small wild fish into and out of large ponds during their filling up and draining.[18] These fish may carry the virus into the water course supplying the farm, and a typical endemic situation is therefore unavoidable.

If carp exposed to infection are still feeding, oral delivery of broad spectrum antibiotics can somewhat reduce the mortality rates at temperatures above 15°C by controlling secondary bacterial infections. Antiviral compounds have not been tested for chemotherapy. The use of oral chemotherapeutants is limited because of the poor feeding response of carp immediately prior to and during the disease outbreaks due to low temperatures. Administration by bathing is limited by the large volume of ponds.

THE STRATEGY OF VACCINATION

As shown in Fig. 16.2, the timing of vaccination and the applicability of methods for vaccine delivery are limited by seasonal phases in carp culture technology and fluctuations of temperature.

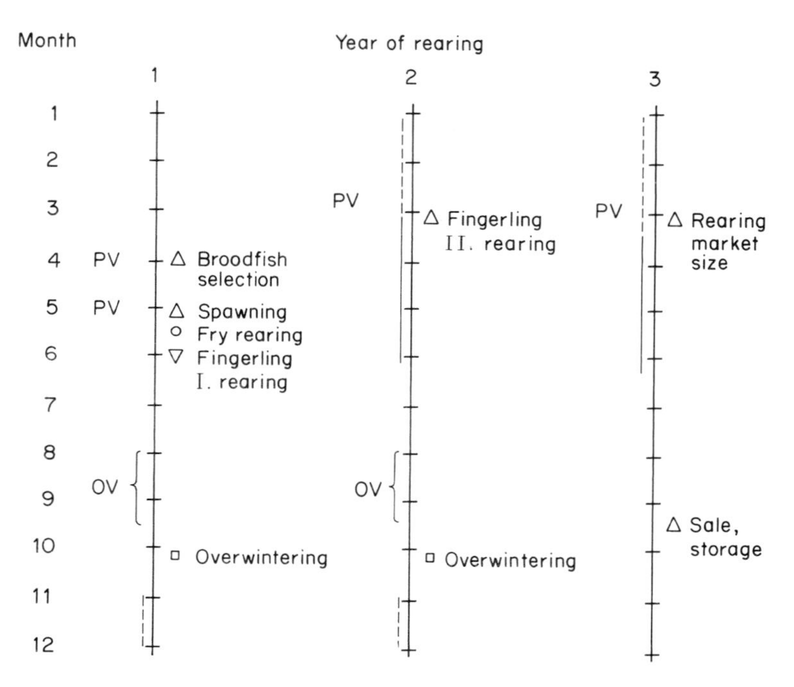

Fig. 16.2 Production phases and handling of carp, seasonality of SVC and timing of vaccination in the three growing seasons cycle (from Fijan[17]). △ = Fishing out and stocking into another pond always practised; ○ = fry rearing for 1 month, end weight 0.2–0.8 g; ▽ = fingerlings are reared by one of the following procedures; (a) stocking with larvae, (b) stocking with 10–14-day old fry, or (c) stocking with 1-month old fry; □ = in some countries fingerlings overwinter in the same pond while in others they are fished out and transferred to special wintering ponds; ¦ = season of infrequent SVC; | = SVC is frequent: PV = intraperitoneal vaccination; OV = oral vaccination.

Depending upon the local climate, the production technology in SVC areas lasts 2 or 3 years. Figure 2 depicts the situation in the 3-year cycle, but phases in production and possibilities for vaccination are essentially the same in the 2-year cycle in southeastern Europe and Japan. It can be seen that, with the exception of fry and broodfish, carp are normally not removed from ponds during water temperatures suitable for immunization.

A practical vaccination strategy would be as follows: broodfish could be vaccinated intraperitoneally (i.p.) preferably immediately after spawning. In order to protect the endangered serologically negative farms from disease introduction, 40–60 days old carp should be vaccinated by bath or orally.

On serologically positive farms, oral or bath delivery of a vaccine should be carried out 1–2 weeks before the water temperature falls below 19–20°C in late summer or at the beginning of autumn. This would be the simplest vaccination procedure for achieving protection of fingerlings. The evoked immunity should last until the increase of temperature in spring above 16–17°C, i.e. in Europe for about 8 months. Ideally, the vaccination procedure should encompass all the fingerlings reared on the farm.

The price of two seasons old carp for stocking makes the i.p. delivery of a vaccine in spring economically feasible, at least in some countries. Such vaccination would be very attractive if vaccines against some other major diseases (e.g. carp erythrodermatitis, columnaris disease, bacterial gill disease, *Pseudomonas* septicaemia) could be delivered simultaneously.

All species cultivated together with carp on a farm should be vaccinated. This prerequisite can be met by bath delivery and by injection of vaccines, but not always by the oral route: predatory fish like sheatfish and pike as well as the planktonophagous silver carp do not take the usual commercially prepared food.

The use of a live vaccine would probably be best during warm weather (to facilitate viral replication) and could be delivered by bath or, on small farms or ponds, by injection. The ideal live vaccine for delivery by the oral route, injection or bath should be highly immunogenic and apathogenic over the whole temperature range in which disease occurs. The efficient dose of virus per fish should be expected to be between 10^4 and 10^5 infective particles. In areas where SVC is endemic the high percentage of serologically positive farms may become a justification for the use of live vaccines, even if the induction of carrier status by them cannot be avoided.

A custom-made spectrum of virus subtypes should be available for formulating vaccines for different regions and locations. For oral delivery, a special food ensuring stability of the virus and its protection from the action of digestive enzymes would be desirable and would minimize the immunogenic virus dose per fish.

NATURE OF VACCINES

The only commercial vaccine is produced by Bioveta, ČSSR, since 1981. It contains two inactivated strains (deposited as CAMP V-236 and V-237) of virus.[9] Details of the strains and production of the vaccine are not available.

Vaccine contains approximately 10^4 $TCID_{50}$ ml.[10] The dose of vaccine in oil emulsion for i.p. injection in spring is 0.5 ml for carp up to 0.5 kg and 1 ml for larger fish.

All other vaccines have been produced for experimental purposes only. Matašin *et al.*[11] prepared an inactivated vaccine from the reference strain of SVCV by adsorption of the virus harvested from EPC cells, onto aluminium hydroxide and inactivation with 0.5 per cent formalin.

The live vaccine tested by Fijan *et al.* (Ref. 12 and unpublished) contained one strain passed on human diploid, BHK-21 and FHM cells. It was not pathogenic for carp at temperatures above 18°C. Fish were vaccinated i.p. and orally on farms at temperatures above 19–20°C. The dose for i.p. vaccination was 6.3×10^4 $TCID_{50}$ per fish of 40–100 g. For oral delivery about 3.1×10^5 $TCID_{50}$ of virus was added for each fish to the food containing 30 per cent protein and delivered during two consecutive days as a dough, stable in water for 4 h.

Kölbl[13] passed two strains on chicken fibroblasts, BHK-21 and EPC cells and selected virions producing large plaques for experimental vaccines. Passages in fish cells reduced both the pathogenicity and immunogenicity, while passages in cells from warm-blooded animals reduced only the pathogenicity. Vaccines were delivered orally in a food with soybean flour. Doses were not specified.

Fijan and Matašin[14] used live unattenuated virus for bath immunization (1×10^5 $TCID_{50}$/ml for 1 h at 21.5°C) of small groups of carp (8 g) in the laboratory.

TESTING AND EFFECTIVENESS OF VACCINES

Injectable vaccines

The Czechoslovak commercial inactivated vaccine has been tested only in the field. Over several consecutive years, large populations of carp were vaccinated i.p. in spring. Percentages of total losses determined at harvest in autumn were lower in vaccinated populations and the vaccination was economically efficient.[10] The small samples of vaccinated fish analysed for serum neutralizing antibodies were positive. In one instance, SVC broke out after vaccination. The vaccination was not technically feasible at water temperatures above 12°C: the fishing out of carp from large ponds is unsafe at such temperatures. Resistance to SVC developed after 1 week at low temperature but the mechanism has not been investigated.

The inactivated vaccine of Matašin *et al.*[11] induced formation of virus neutralizing antibodies (1:8 to 1:32) in over 50 per cent of carp kept in the laboratory at 13°C for 1 month after i.p inoculation, but mortality rates after challenge were the same in vaccinated and control fish. In laboratory tests, addition of three different antibiotics to the vaccine did not influence mortalities after challenge. In field trials, kanamycin was added to the vaccine. Carp for stocking of two ponds on two farms were injected in spring at 11–13°C with 1.4×10^4 $TCID_{50}$ of inactivated virus. It was impossible to assess the specific efficacy of this vaccination since clinical SVC was not detected on these farms and fish vaccinated in the field were not challenged in the laboratory. However, losses were lower in injected populations: 57 per cent versus the average of 70 per cent at one and 20 per cent versus the average of 60 per cent at the other farm. These differences can be partially attributed to the injection of the antibiotic in the vaccine. More work is needed to clarify the discrepancy in results with inactivated vaccines: development of virus neutralizing antibody was not accompanied by protective immunity in the laboratory, while field vaccinations decreased losses on farms.

Bath vaccines

Bath vaccination with the live unattenuated virus vaccine, used by Fijan and Matašin[14] produced good protection in young carp when challenged by bath or i.p. injection 1 month after vaccination. In other studies, Hill[15] and Ahne[16] obtained similar results.

Oral vaccines

The pilot i.p. and oral vaccinations of carp in the field using a live vaccine at the end of summer (Ref. 12 and unpublished) evoked a long lasting protective immunity. Groups of vaccinated and control carp were challenged in the laboratory by i.p. inoculations 2, 3 and 9 months after vaccination. The total mortalities in these three tests were 55 per cent for 40 control fish, 22 per cent for 60 orally vaccinated and 5 per cent for 60 i.p. vaccinated carp. Neutralizing antibodies (1:8 to 1:16) were present 1 and 2 months after vaccination: 7.5 per cent of 118 sera from orally vaccinated and 38.2 per cent from i.p. vaccinated carp were positive. In the control group, neutralizing antibody was detected only in the sample taken after 2 months: 23 per cent of 30 sera were positive. Sera taken before the vaccination and those after 3 and 9 months were all negative.

In large-scale oral vaccination trials of the live vaccine in three ponds on different farms (Fijan *et al.*, unpublished), neutralizing antibodies were not detected. One vaccinated and one control population were tested in the laboratory for resistance to i.p. challenge 2, 6 and 7.5 months after vaccination. Total mortality over the three tests was 86 per cent of 80 control carp and 36 per cent of 80 vaccinated carp. The following spring, vaccinated fish were stocked into several grower ponds and in three instances mixed with unvaccinated stock. In one such pond with a mixed population, SVC caused slight losses among unvaccinated carp. The disease did not appear in any other pond of these farms.

In large-scale laboratory trials of oral immunization, Kölbl[13] imitated seasonal temperature oscillations in ponds. The best resistance to challenge developed after simultaneous administration of live avirulent and a partially attenuated strain, which were also used in field vaccinations. Mortality rates in pond experiments and laboratory challenges of small samples of carp demonstrated good potency of these and other vaccines.

While the above trials were reliable enough to demonstrate a protective effect of most vaccine preparations accurate interpretation of data is difficult. The existing health status of carp at the time of vaccination has been recognized as important. Kölbl[13] found potency of vaccination was poor in carp populations affected by intestinal tapeworms or blood flagellates. Interference with the development of resistance in carp affected by bothriocephalosis after injection with live partially attenuated virus was also encountered in the first pilot field experiment by Fijan *et al.* (unpublished). Furthermore, field trials have often been hampered by inadequate natural challenge and laboratory challenges of carp vaccinated in the field were often carried out on too small samples due to the difficulty in catching enough fish during late autumn and winter. Adequate samples can often be obtained only at the final harvesting of fish from ponds. However, workers have no alternative method for vaccine testing, the only reliable test for potency being challenge with the SVC virus as neutralizing antibody titres do not seem to correlate well with protection.

LIMITATIONS OF PRESENT VACCINES

The few vaccines tested so far have some obvious limitations but further, more reliable tests, under wider ranging conditions are necessary to assess their usefulness. Lack of adequate information on production, composition and potency of the first commercially produced inactivated vaccine[9] presently precludes recommendation for use on a wide scale. The inactivated preparation of Matašin *et al.*[11] lacks potency at low temperature and was not tested at higher ones. The quantity of antigen in it was low and different doses have not yet been tested.

As expected from the data on host–virus interactions, most live vaccines delivered at nearly optimum temperatures for development of resistance had an adequate potency but their safety has not been tested properly. The partially attenuated vaccine of Fijan *et al.* (Ref. 12 and unpublished) was not safe when unvaccinated fish were stocked together with vaccinated ones: one SVC outbreak in unvaccinated carp was suspected to be caused by shedding of vaccinal virus. Inefficiency of vaccination due to the inadequate health status of fish is a risk inherent to carp farming.

OVERCOMING LIMITATIONS

Improvement of the present status of SVC vaccination requires consideration of management changes in pond farming as well as further research on the disease and vaccine formulation, delivery and testing.

Since the practical limitations for vaccinating carp presented in Fig. 16.2 cannot be easily altered in classical pond farming, a strategy for separate production of vaccinated fish for stocking the weather-dependent farms could be considered. Such carp could be reared in heated effluents, thermal waters or in recirculation units and, 1 month prior to stocking pond farms in spring, vaccinated by bath at above 20°C in an inactivated vaccine. Small ponds on some farms could be fished out in late summer and vaccinated by the same method.

On classical farms, a more complex fish health protection approach, including vaccination with multivalent vaccines against major diseases, should be aimed at reducing the often encountered high losses. The attractiveness of a SVC or any other vaccine would thus be increased.

The next phase of research on SVC vaccination needs models for vaccination trials in the laboratory and on farms, as well as standards for vaccine potency testing. Virus serotypes, their geographic range and their resistance-inducing components need to be defined. If the oral delivery remains the principal choice of immunoprophylaxis on large farms, safe and highly immunogenic attenuated strains will be needed. A synthetic vaccine composed of non-infectious viral antigens as well as a carrier and adjuvants for its bath and i.p. delivery would be desirable. Unfortunately the number of researchers and availability of funds for attaining these goals are presently very limited.

COMMERCIAL PROSPECTS

None of the preparations so far tested is suitable for commercial production and one can only speculate about more adequate future candidates.

A live vaccine for oral delivery in autumn would be attractive for use in endemic areas on serologically positive farms and would probably require the least costly research and development. However, if a carrier status resulted from its use, many countries would not license it and some would impose serious restrictions to its application.

The husbandry prerequisites for delivery of inactivated vaccines by bath do not presently exist on large farms, though if such a vaccine was available, especially if combined with vaccines to other disease agents, it may stimulate the profound changes in carp production techniques necessary to incorporate its use. An inactivated vaccine could also be used for vaccination of broodfish.

CONCLUSIONS

SVC is one of the principal causes of losses on carp farms in moderate climates and immunoprophylaxis offers the only effective means of controlling this widespread disease.

The relatively limited, mostly practically oriented research has demonstrated the possibility of inducing a sufficiently effective and long lasting resistance in carp by live vaccines. Experiments with other susceptible species kept in ponds together with carp have not been carried out yet.

Contrary to the intensive tank or cage husbandry of salmonids, the extensive pond farming of cyprinid fishes imposes considerable restrictions on the timing and modes of vaccine delivery. An additional problem represents the frequent occurrence of other diseases particularly parasites, which can hamper the development of specific resistance.

With regard to the most applicable methods on carp farms, oral vaccination with attenuated virus in autumn and the intraperitoneal injection of inactivated virus in spring have been tested, but only the former method proved to be partially successful. The development of procedures for the commercial production of an attenuated vaccine for oral or bath delivery, as well as of an inactivated or synthetic vaccine for bath or injection delivery will have to be preceded by more research. Given this, it is unlikely that a commercial SVC vaccine will be available in the very near future.

REFERENCES

1 Fijan, N., Petrinec, Z., Sulimanović, D. and Zwillenberg, L. O. (1971). Isolation of viral causative agent from the acute form of infectious dropsy of carp. *Vet. Arh.*, **41**, 125–138.

2 Kölbl, O. and Kainz, E. (1977). Die Verbreitung des Erregers der Infektiösen Bauchwassersucht bzw. der Frühlingsvirämie inden österreichischen Karpfenteichwirtschaften. *Österreichs Fischerei*, **30**, 80–83.

3 Wizigmann, G., Dangschat, H., Baath, Ch. and Pfeil-Putzien, C. (1983). Untersuchungen über Virusinfektionen bei Süsswasserfischen in Bayern. *Tierärztl. Umschau*, **38**, 250–258.

4 Hattenberger-Baudouy, A. M. and de Kinkelin, P. (1988). Diagnostic of virus diseases of fish with special reference to cyprinid virus infections. In: *Ichthyopathology in Aquaculture*, eds. N. Fijan, S. Cvetnić and T. Wikerhauser, JAZU, Zagreb (in press).

5 Ahne, W. (1985). *Argulus foliaceus* and *Piscicola geometra* as mechanical vectors of spring viremia of carp virus (SVCV). *J. Fish Dis.*, **8**, 241–245.
6 Ahne, W. (1982). Vergleichende Untersuchungen über die Stabilität von vier fischpathogenen Viren (VHSV, PFR, SVCV, IPNV). *Zbl. Vet. Med. B.*, **29**, 457–476.
7 Békesi, L. and Csontos, L. (1985). Isolation of spring viremia of carp virus from asymptomatic broodstock carp, *Cyprinus carpio* L. *J. Fish Dis.*, **8**, 471–472.
8 Ahne, W. (1986). Unterschiedliche biologische Eigenschaften 4 cyprinidenpathogener Rhabdovirusisolate. *J. Vet. Med. B.*, **33**, 253–259.
9 Tesarčik, J. and Macura, B. (1981). Field carp vaccination against spring viremia on the fish farms of the State Fishery. *Bul. VÚRH Vodnany*, **17**, 3–11 (in Czech.).
10 Tesarčik, J. and Macura, B. (1988). Spring viremia of carp – development of a vaccine in Czechoslovakia. In: *Ichthyopathology in Aquaculture*, eds. N. Fijan, S. Cvetnić and T. Wikerhauser, JAZU, Zagreb (in press).
11 Matašin, Ž., Fijan, N. and Petrinec, Z. (1988). Vaccination of carp against spring viremia with an inactivated vaccine. In: *Ichthyopathology in Aquaculture*, eds. N. Fijan, S. Cvetnić and T. Wikerhauser, JAZU, Zagreb (in press).
12 Fijan, N. Petrinec, Z., Stancl, Z., Kezic, N. and Teskeredzic, E. (1977). Vaccination of carp against spring viremia: comparison of intraperitoneal and peroral application of live virus to fish kept in ponds. *Bull. Off. int. Epiz.*, **87**, 441–442.
13 Kölbl, O. (1980). Diagnostic de la viremie printaniere de la carpe et essais d'immunisation contre cette maladie. *Bull. Off. int. Epiz.*, **92**, 1055–1068.
14 Fijan, N. and Matašin, Ž. (1980). Spring viremia of carp: preliminary experiments on vaccination by exposure to virus in water. *Vet. Arh.*, **50**, 215–220.
15 Hill, B. J. (1977). Studies on SVC virulence and immunization. *Bull. Off. int. Epiz.*, **87**, 455–456.
16 Ahne, W. (1980). *Rhabdovirus carpio* – Infektion beim Karpfen (*Cyprinus carpio*): Untersuchungen über Reaktionen des Wirtorganismus. *Fortschritte der Veterinärmedizin*, **30**, 180–183.
17 Fijan, N. (1988). Viral diseases of cyprinids in European pond culture. In: *Ichthyopathology in Aquaculture*, eds. N. Fijan, S. Cvetnić and T. Wikerhauser, JAZU Zagreb (in press).
18 Fijan, N. (1984). Vaccination of fish in European pond culture: prospects and constraints. *Symposia Biologica Hungarica*, **23**, 233–241.
19 Fijan, N., Matašin, Ž., Jeney, A., Olah, J. and Zwillenberg, L. O. (1984). Isolation of *Rhabdovirus carpio* from sheatfish (*Silurus glanis*) fry. *Symposia Biologica Hungarica*, **23**, 17–24.
20 Hattenberger-Baudouy, A. M., Danton, M. and Merle, G. (1987). Infection experimentale de l'alevin de carpe *Cyprinus carpio* L. par le virus de la viremie printanière de la carpe (V.P.C.) en eau chaude. *Bull. Fr. Pêche Piscic.* **307**, 89–90.

Note added in proof.

Carp less than one month old (which are immunologically immature) can be affected by SVC at high temperatures. The sheatfish fry mortality at 22°C[19] and carp fry mortalities at this[20] and higher temperatures (Fijan *et al.*, unpublished) suggest that young fish cannot produce interferon and/or neutralizing antibodies. However, SVC outbreaks seem to be extremely rare in this very susceptible age group. This epidemiological pattern is not clear.

17

Vaccination against Channel Catfish Virus

J. A. Plumb

THE DISEASE PROBLEM

Channel catfish virus (CCV) is an acute highly infectious disease of juvenile channel catfish, *Ictalurus punctatus*, that is widely spread throughout the areas where channel catfish are cultured. Although there is only one confirmed case of CCV outside the United States (Honduras), there is great concern about its importation into areas where channel catfish are being introduced.

CCV is caused by a herpesvirus, *Herpesvirus ictaluri*.[1] As is the case with other herpesviruses, CCV is difficult to detect unless an active epizootic is in progress. During epizootics, CCV is readily isolated on either brown bullhead (BB) cells or channel catfish ovary cells (CCO).

CCV does elicit an immune response in channel catfish of all ages, and the detection of CCV neutralizing antibody has been used to identify potential CCV carrier fish.[2] Amend and McDowell[3] and Plumb and Jezek[4] pointed out some advantages, disadvantages, and problems associated with this technique. Also, Bowser and Munson[5] showed that CCV antibody titres in 4-year-old channel catfish that had survived a CCV epizootic were higher from July to October than in January and April. Wise and Boyle[6] and Wise *et al.*[7] utilized a CCV specific DNA probe to detect CCV antigen in tissues of adult channel catfish.

CCV may cause an economically devastating disease as it can cause 90 per cent mortality in fry or fingerlings. The inability to control or eliminate the disease has been partially responsible for the demise of at least two catfish farms in the United States, and it has restricted production on others.

Although the channel catfish is the primary species that is severely affected by CCV, there are differences in susceptibility among strains of catfish (Fig. 17.1).[8] Wild strains are more sensitive to CCV than hatchery strains and apparently crossbred strains are more resistant than parental groups. White catfish, *I. nebulosus* and blue catfish, *I. furcatus*, are refractory to infection except under experimental conditions. Plumb *et al.*[9] attempted to infect the

FISH VACCINATION
ISBN 0-12-237485-1

European catfish, *Silurus glannis,* and concluded that this species is also refractory to CCV. Thus, even if CCV were introduced into Europe, the European catfish may not be adversely affected. The walking catfish, *Clarias batrachus*, is also refractory to CCV.[10]

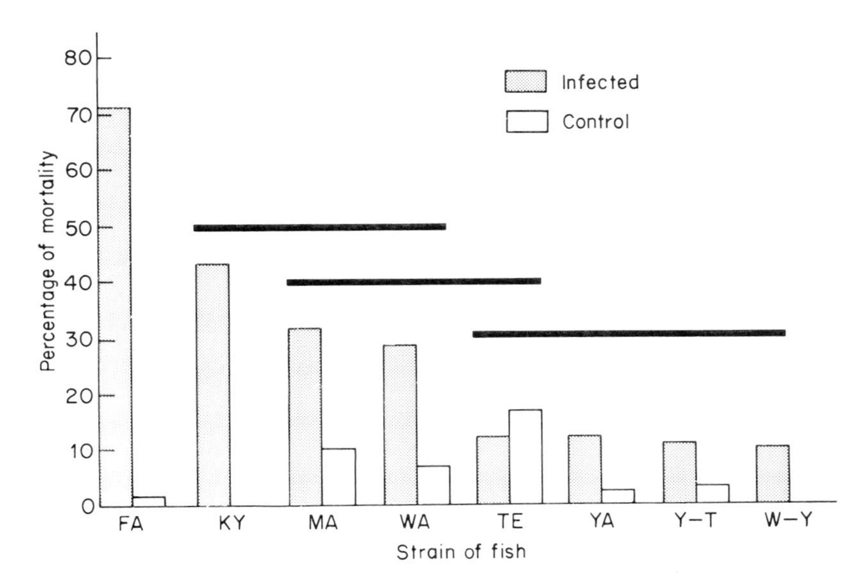

Fig. 17.1 Mortality of different strains of channel catfish fed channel catfish virus at 1–3 months of age. The catfish strains infected were: Falcon (FA), Kentucky (KY), Marion (MA), Tennessee (TE), Warrior (WA), Yazoo (YA), Warrior × Yazoo (W–Y), Yazoo × Tennessee (Y–T). The means of any two strains that do not fall under the same line are significantly different ($Q_{0.05}$). (From Plumb *et al.*[8])

CCV causes problems in channel catfish between 1 week and 6 months of age.[11] Only occasionally has CCV been isolated from diseased fish older than 6 months; for instance, Bowser *et al.*[12] isolated the virus from adult catfish in January. Most reports of CCV disease have been during the warm summer or early fall months. Two factors occur simultaneously during the summer that contribute to high mortalities: (1) there is an abundance of susceptible age and size fish; and (2) the water temperature is in the optimum range. The highest mortality (90 to 100 per cent) occurs when water temperature is 25°C to 34°C; temperatures in this range are common in the channel catfish growing region of the United States from June to September. At temperatures of 18°C to 25°C, mortalities range from 40 to 70 per cent; below 18°C, few fish die.

Although records are not good, the disease appears to be more prevalent in heavily stocked fingerling ponds. Poor water quality, such as low oxygen concentration is also associated with high incidence of CCV and high mortality rates, however, a causal relationship has not been proven experimentally. Secondary bacterial infections of *Flexibactor columnaris* and *Aeromonas hydrophila* often occur during and after CCV epizootics, which can extend the mortality beyond the 2 to 3 week period normally found when CCV alone is involved.

CCV infects fish in ponds, raceways or holding tanks. In tanks, mortality as high as 100 per cent has occurred. In ponds mortalities up to 95 per cent have been reported, although 40 to 60 per cent is more common, and mortality as low as 5 per cent has been reported by fish farmers.

CCV is transmitted from infected to non-infected fingerlings through the water by cohabitation. The virus can be transmitted by feeding contaminated feeds; therefore, it can be assumed that CCV can be contracted by cannibalizing dead or moribund fish. The virus survives for over 14 days in infected fish held on ice, but CCV does not survive in decomposing fish held at 23°C for more than 1.5 days.[13] As long as the fish are chilled they can be a source of infectious CCV for days, but not in ponds where fish can decompose rapidly and are a source of infectious virus for only a brief period.

Evidence for vertical transmission is circumstantial in that CCV often reoccurs in offspring from certain groups of broodfish in successive years and the disease occurs in very young fish held in water from wells and springs that harbour no wild fish. Isolation of CCV from adult fish makes the possibility of adult carrier fish and vertical transmission feasible.[12] Also, Plumb *et al.*[14] showed the presence of CCV antigen in ovarian tissues of experimentally infected adult channel catfish using indirect immunofluoresence. Numerous attempts have been made to demonstrate vertical transmission via reproductive products, but all attempts to prove this point have failed.

At the present time, it is not known if all CCV isolates are serologically homogeneous. Sufficient antigenic and serological data are not available to make that determination. However, with respect to pathogenesis, histopathology, and cytopathology, the CCV isolates appear homogeneous.

Therapeutic control of CCV disease does not exist, although many farmers feed medicated feed to prevent secondary bacterial infections. Since the disease is temperature dependent, Plumb[15] suggested that reduction of water temperature to 18°C would arrest mortality, but this must be done within 24 h of infection. This practice is limited to facilities that have access to cool water, and it has little application to the catfish industry.

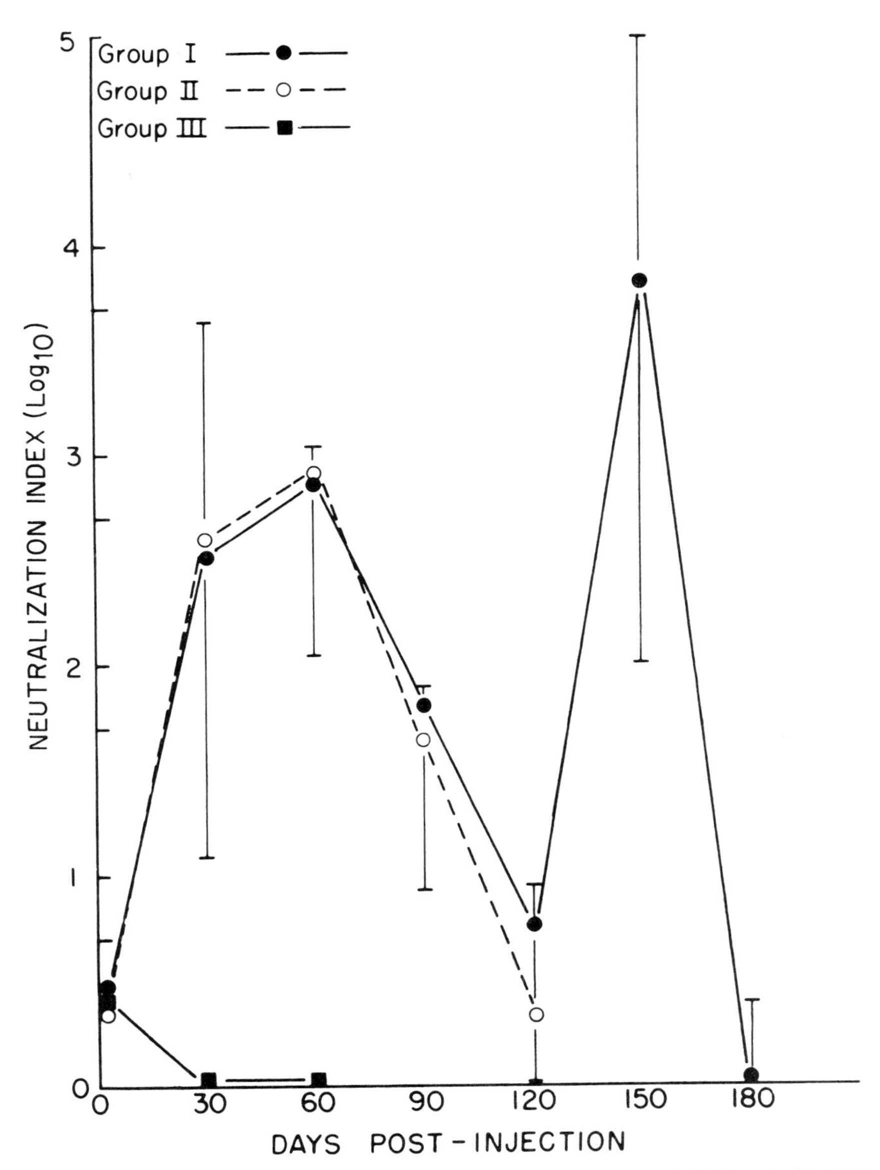

Fig. 17.2 Neutralization indices (log base 10) of 2-year-old channel catfish injected with CCV. Groups I and II were inoculated with infectious virus and Group III was inoculated with heat killed virus. Group I was given a booster injection 120 days after the initial inoculum. (From Plumb.[2])

THE IDEAL VACCINE AND ITS STRATEGIC USE

The ideal vaccine against CCV must be effectively applied in a bath. This is necessary due to the small size and young age of the most susceptible fish and the vaccine should be immunogenic to channel catfish of 1–10 cm length. Application of the vaccine would be made at the earliest age and size possible and would allow a large number of fish to be vaccinated in a small volume of vaccine as fish are moved from hatching troughs to the fingerling ponds. This could be done with very little additional handling of the fish.

NATURE OF VACCINES

Ideally, a CCV vaccine would be composed of killed virus or an attenuated preparation. Unfortunately, killed CCV has very low immunogenicity (Fig. 17.2).[2] A vaccine described by Noga and Hartman[10] was made with an attenuated virus. CCV was passaged repeatedly in walking catfish, *Clarias batrachus*, kidney cells (KIK), and after 60 passages the virulence of the virus was eliminated and the attenuated CCV was designated V60. Virus in V60 vaccine developed plaques on BB cells that averaged 0.38 mm in diameter as compared to an average plaque diameter of virulent virus of 1.16 mm when incubated at 29°C. Virus release from cells was similar for the V60 and the virulent CCV. The mode of delivery of V60 was by immersion of fish in the vaccine, or by intraperitoneal injection.

VACCINE TRIALS

Noga and Hartmann[10] described virulence and vaccination trials of the V60 vaccine where fingerling catfish were injected intraperitoneally with three types of CCV (Table 17.1).

Table 17.1 *Survival of channel catfish injected with attenuated and unattenuated CCV and then challenged with virulent CCV-3 3 weeks later* (from Noga and Hartmann[10])

Virus strain injected	% Survival after initial injection	% Survival after challenge (3 weeks after vaccination)[c]
HBSS Control	100	0
V60[a]	100	90
CCV-5[b]	0	—
CCV-3[c]	20	—

[a]V60: CCV attenuated by 60 passages in walking catfish kidney cells (KIK).
[b]CCV-5: virulent CCV passaged 35 times in brown bullhead (BB) cells.
[c]CCV-3: virulent wild type CCV passaged three times in BB cells.

Three weeks after injection with these preparations, the V60 vaccinated and Hanks balanced salt solution (HBSS) injected control fish were challenged with the CCV-3 (wild type). In the V60 vaccinated fish, 90 per cent survived compared to 0 per cent survival in the HBSS control fish (Table 17.1). Thus, the attenuated V60 vaccine was highly protective.

Immersion vaccination of 1 to 2 month old channel catfish using the V60 attenuated CCV vaccine was described by Walczak *et al.*[16] (Table 17.2). The fish were first placed in 8 per cent NaCl for 30–60 s followed by immersion for 30 min in a solution containing 6000 plaque forming units (pfu)/ml of the V60 virus. After 3 weeks, 36 immunized fish were challenged with virulent wild type CCV, while 54 fish were again immunized as described. The fish vaccinated twice were then challenged 3 weeks following the booster vaccination. Booster vaccinated fish were more effectively protected than those receiving a single immersion vaccination.

The Relative Per cent Survival (RPS)[17] for single immersed fish was 23.9 per cent, compared to a RPS of 85 per cent for the booster immersed group.

Table 17.2 *Summary of survival of channel catfish immersion vaccinated with an attenuated strain of CCV and challenged with virulent CCV (Walczak* et al.[16])

Method of vaccination	Per cent survival following challenge		Relative per cent survival[a]
	vaccinated	non-vaccinated	
Single immersion	30	8	23.9
Booster immersion	97	20	85

[a]Relative per cent survival calculated according to the following formula of Amend[17]

$$\text{Relative per cent survival} = 1 - \frac{\text{\% mortality vaccinates}}{\text{\% mortality non-vaccinates}} \times 100$$

LIMITATIONS OF PRESENT VACCINE

The present CCV vaccine is limited to its use on fingerling channel catfish and the range of age or size of fish on which it is effective is yet unknown. It would be advantageous to vaccinate young fish as soon as they are immunocompetent so that the maximum number possible could be immunized with the least effort. The minimum age of immunocompetency towards CCV vaccine is not yet known. Antigens of CCV appear to be heat labile (Fig. 17.2), excluding the use of heat-killed vaccines. Attenuated vaccines always present the risks of reversion to a virulent form, therefore the CCV vaccine should be used only where water from vaccinated fish cannot come in contact with non-vaccinated stocks. Whether or not vaccinated fish become carriers of CCV is not known.

COMMERCIAL PROSPECTS

The possibilities of developing a commercial attenuated vaccine for CCV to be used in channel catfish are limited. First, it is unlikely that the demand for a large volume of vaccine would occur. CCV disease is seasonal and related to age of the fish and environmental conditions; therefore, it can often be dealt with through appropriate management. However, farms that have recurring large losses of fry or fingerlings would welcome an effective vaccine against CCV.

CONCLUSIONS

There is little optimism that an effective and economical CCV vaccine will soon be available. More experimental work needs to be done before an effective and economical CCV vaccine is available. The limited market for a vaccine makes it unlikely that a commercial vaccine producer would invest in such research. However, it has been demonstrated that an attenuated CCV vaccine is an effective immunogen when administered by intraperitoneal injection or by immersion.

REFERENCES

1 Wolf, K. and Darlington, R. W. (1971). Channel catfish virus: a new herpes-virus of ictalurid fish. *J. Virol.* **8** (4), 525–533.

2 Plumb, J. A. (1973). Neutralization of channel catfish virus by serum of channel catfish. *J. Wildl. Dis.*, **9** (4), 324–330.

3 Amend, D. F. and McDowell, T. (1984). Comparison of various procedures to detect neutralizing antibody to the channel catfish virus in California brood channel catfish. *Prog. Fish-Cult.*, **46** (1), 6–12.

4 Plumb, J. A. and Jezek, D. A. (1983). Channel catfish virus disease. In: *Antigens of Fish Pathogens: Development and Production for Vaccines and Serdiagonostics*, ed. D. P. Anderson, M. Dorson and Ph. Dubourget, Lyon: Collection Marcel Merieux, pp. 33–50.

5 Bowser, P. R. and Munson, A. D. (1986). Seasonal variation in channel catfish virus antibody titers in adult channel catfish. *Prog. Fish-Cult.*, **48** (3), 198–199.

6 Wise, J. A. and Boyle, J. A. (1985). Detection of channel catfish virus in channel catfish, *Ictalurus punctatus* (Rafinesque): use of a nucleic acid probe. *J. Fish Dis.*, **8**, 417–424.
7 Wise, J. A., Bowser, P. R. and Boyle, J. A. (1985). Detection of channel catfish virus in asymptomatic adult catfish, *Ictalurus punctatus* (Rafinesque). *J. Fish Dis.*, **8**, 485–493.
8 Plumb, J. A., Green, O. L., Smitherman, R. O. and Pardue, G. B. (1975). Channel catfish virus experiments with different strains of channel catfish. *Trans. Am. Fish. Soc.*, **104** (1), 140–143.
9 Plumb, J. A., Hilge, V. and Quinlan, E. E. (1985). Resistance of the European catfish (*Silurus glanis*) to channel catfish virus. *J. Applied Ichthyology*, **1**, 87–89.
10 Noga, E. J. and Hartmann, J. X. (1981). Establishment of walking catfish (*Clarias batrachus*) cell lines and development of a channel catfish (*Ictalurus punctatus*) virus vaccine. *Can. J. Fish. Aquat. Sci.*, **38**, 925–930.
11 Plumb, J. A. (1978). Epizootiology of channel catfish virus disease. *Marine Fish. Rev.*, **49** (3), 26–29.
12 Bowser, P. R., Munson, A. D., Jarboe, H. H., Francis-Floyd, R. and Waterstrat, R. P. (1985). Isolation of channel catfish virus from channel catfish *Ictalurus punctatus* (Rafinesque) broodstock. *J. Fish Dis.*, **8**, 557–561.
13 Plumb, J. A., Wright, L. D. and Jones, V. L. (1973). Survival of channel catfish virus in chilled, frozen, and decomposing channel catfish. *Prog. Fish-Cult.*, **35** (3), 170–172.
14 Plumb, J. A., Thune, R. L. and Klesius, P. H. (1981). Detection of channel catfish virus in adult fish. *Develop. Biol. Stand.*, **49**, 29–34.
15 Plumb, J. A. (1972). Effects of temperature on mortality of fingerling channel catfish (*Ictalurus punctatus*) experimentally infected with channel catfish virus. *J. Fish. Res. Board Can.*, **30** (4), 568–570.
16 Walczak, E., Noga, J. and Hartmann, J. X. (1981). Properties of a vaccine for channel catfish virus disease and a method of administration. *Develop. Biol. Stand.*, **49**, 419–429.
17 Amend, D. F. (1981). Potency testing of fish vaccines. *Develop. Biol. Stand.*, **49**, 447–454.

18

Vaccination against Protozoan and Helminth Parasites of Fish

G. Houghton, R. A. Matthews and J. E. Harris

INTRODUCTION

A wide variety of invertebrate parasites infect both freshwater and marine fishes. They belong to a number of distinct groups Protozoa, Platyhelminthes, Nematoda, Acanthocephala, Mollusca, Annelida and Arthropoda. A subtle equilibrium, both in terms of host specificity and damage to the host, often exists between the host and the parasite particularly in wild fish populations.[1] This is often considered to be a result of a sophisticated adaptation which has developed over a long evolutionary period. Serious disease epidemics and mortalities, as a result of invertebrate parasites, are often only observed when conditions for the fish host are less than ideal and this is seen most frequently during adverse environmental conditions, or when stocking densities in natural fisheries are high or in modern intensive culture systems. The parasitic protozoa and helminths have received the most recent attention.

Much of the early work on these groups has been necessarily concentrated on morphology, taxonomy, life-cycles and transmission rather than on the immunological relationships between the host and parasite and any subsequent pathology. The relatively few early studies of the immune response of fish to their parasite fauna have been reviewed by Lom[2] and Snieszko;[3] the former being specifically concerned with protozoan and the latter with metazoan parasites. More recently, Kennedy[1,4] and van Muiswinkel and Jagt[5] have produced reviews.

Knowledge of the immunological relationships between fish and their parasites is small in comparison with that concerning bacteria and viruses. There are numerous reasons for this, the most significant being:

1 the problems encountered in laboratory manipulation of protozoan and helminth parasites
2 the difficulties of *in vitro* and *in vivo* culture of the organisms
3 the intrinsic complexities of the parasites and their complex life-cycles, often involving several hosts.

FISH VACCINATION
ISBN 0-12-237485-1

A comparable situation exists in both veterinary and clinical medicine, where the development of vaccines to protozoan diseases (malaria, trypanosomiasis) and to helminth diseases (schistosomiasis, fascioliasis), although of tremendous potential value, has yet to be satisfactorily accomplished. Furthermore in fish, in terms of morbidity and mortality, fewer problems are caused by helminth and protozoan diseases than by viruses or bacteria.

Nevertheless, advances have recently been made and in this review, examples of some of the more economically important protozoan and helminth diseases will be given where some progress has been made in investigating the immunological relationships between host and parasite and in the development of vaccines.

ICHTHYOPHTHIRIASIS

Epidemiology and Aetiology

'Ich' or white-spot is a major disease and one of the most common parasites of freshwater fishes. It has a world wide distribution from tropical to subarctic zones.[6] One reason for its universal distribution is that most freshwater fish can serve as its host, and there does not appear to be any species with any natural resistance. In wild or natural habitats, enzootic subclinical infections are widespread (thus forming a reservoir of infection able to spread into farms on the same water system) but epizootics have been reported most often in fish farms or hatcheries where, under suitable ambient circumstances, consequent morbidity or mortality may occur leading to significant economic losses.

Epizootics of the disease in carp have been described as early as the 10th century by Su-Shih.[7] The parasite is known to cause severe epizootics not only in food fish, including trout, salmon, catfish, eel, carps and tilapia, but also in valuable ornamental species. Indeed it has been suggested that 'Ich' was introduced into many countries with imported aquarium goldfish (*Carassius auratus*) and inadvertent introduction is a common source of the parasite.

The causative agent is a holotrichous ciliate, *Ichthyophthirius multifiliis* Fouquet, 1876. Nigrelli *et al.*,[6] have postulated that populations of *I. multifiliis* of diverse geographic regions may differ but a comprehensive study of parasites from a variety of sources by Ventura and Paperna[8] indicates that this is not so. The infective stages (theronts – Fig. 18.1) invade the epithelium of the gills and skin affecting the respiratory, excretory and protective functions of these organs.[9,10] The epidermis responds proliferatively to produce the typical grey–white lesions. The mature trophozoite, or trophont, leaves the epidermis, swims for a short while, then rests on the bottom of the pond or aquarium where it secretes a cyst. Reproduction by repeated binary fission produces a new generation by infective theronts. Affected fish often 'flash' against the sides or bottom of the aquaria. The fish become lethargic and death commonly ensues. The infection is associated with a broad temperature spectrum below 27°C and the optimal range is from 21–24°C. At these temperatures, the parasites may mature in 3–4 days, but at 10°C, the cycle may require up to 5 weeks. In warmwater fish farms in the USA the disease shows a marked seasonality with the greatest prevalence in mid-winter, although diagnostic records show that the number of acute cases recognized by farmers is highest in April when water temperatures reach an optimum.[11] In Europe it is essentially a spring and summer condition.

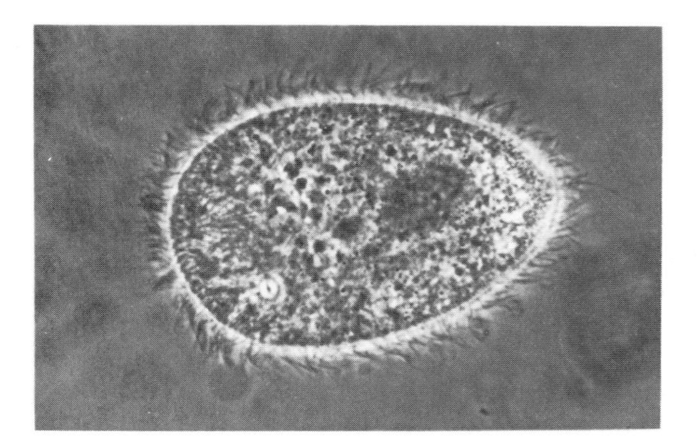

Fig. 18.1 Theront of *Ichthyophthirius multifiliis* (length 34 μm).

Chemotherapy

The mortalities caused by the disease are difficult to control and these difficulties arise from several features of the parasite's life-cycle. Although the infective or invasive stage (theront) is vulnerable to an array of practical control measures, it is short-lived and diverse in the time of its appearance. Within an infected population, this stage can be generated continuously. In contrast, most of its life-cycle is spent within the host epithelium where it is apparently unaffected by chemotherapeutants added to fish feeds or by externally, topically applied methods.[12] Various methods of control of ichthyophthiriasis have been reviewed by Farley and Heckmann.[13] Some methods have been partially successful in treating the free-living forms including increased salinity, methylene blue, mercurochrome, quinine acriflavine, chloramine, nifurpurinol and pyrimidyl mercuro acetate. Commercially, however, they are not viable due to either their cost, toxicity or ineffectiveness. The most successful treatment is bathing in formalin (at approximately 200 ppm for 1 h daily or 25 ppm continuous exposure; this kills both the free-swimming trophozoite or theront, higher doses being necessary for the treatment of cysts) or malachite green in doses ranging from 0.05 ppm to 1 ppm. Malachite green is often used in conjunction with a variety of other chemicals and possibly the most successful is a mixture of malachite green (0.05 ppm) and formalin (15 ppm).[11] Nevertheless, even these are expensive and may be toxic to fish already experiencing stress. Other methods of preventing reinfection, apart from chemotherapy, have involved increased water flow, increased temperature or transfer to clean aquaria.[14] Ascorbic acid in the diet[15] has recently been shown to increase the survival of infected fish.

Immune resistance

More recently, attention has centred around the prospects of preventive immunization as a means of control, not only because of the obvious limitations of chemotherapeutic and other methods, but because of the recognition that a protective immunity develops in previously infected fish. It has long been recognized that fish recovering from an infection possessed varying degrees of resistance to subsequent re-infection and that the level of this immunity could be correlated with the intensity of the primary infection. The duration of resistance is, however, not well established; it may be of short duration, or long lasting, depending on the conditions in which the fish are kept.[16] Protection in adult carp (*Cyprinus carpio*), following exposure to a controlled primary infection, lasted for at least 35 days if fish were kept in an environment free of parasites, but this could be extended for up to 8 months if immune fish were maintained under conditions of continuous exposure to the parasite.[16] In juvenile carp, Houghton[17] found that protection to ichthyophthiriasis extends at least to 3 months following infection with controlled numbers of theronts but, thereafter, the degree of protection decreases. Recently, McCallum[18] has shown that in black mollies, *Poecilia latipinna*, resistance was only maintained for any significant length of time if there were high initial levels of infection. Both carp[19] and salmon[20] have been shown to be refractory to re-infection and Lom[2] showed that at least in carp this was temperature-dependent; those recovering from infection at 26 °C were immune whilst fish kept between 15–20°C could be re-infected.

Subasinghe[21] produced immunity in young carp by exposure to defined numbers of trophozoites on three separate occasions of 14-day intervals; challenge 28 days later resulted in no mortalities. A similar regime was used by Houghton and Matthews[14] to induce immunity in juvenile carp but using a quantitative number of theronts rather than trophozoites, with the immunizing dose consisting of approximately 2000 theronts per fish. A potentially lethal challenge, consisting of approximately 8000 theronts per fish, produced no mortalities (Table 18.1). Protection has also been elicited in catfish by intraperitoneal injection of ground trophozoites with or without the addition of Freund's adjuvant.[22]

Table 18.1 *The effect of a potentially lethal challenge of* I. multifiliis *after immunization with three serial infections of the parasite. Data represent pooled results from two separate experiments* (from Houghton and Matthews[14])

Group and no. of fish	% infected	Degree of infection	% mortality
Immunized (120)	19	very light, periphery of fins	0
Controls (30)	100	very heavy, entire body	100

The mechanisms of resistance are not fully understood, but evidence suggests antibody may be involved. Hines and Spira[16,23] found immunity in mirror carp (*C. carpio* L.) was induced with 200 trophozoites per fish and sera from the 'immune' fish immobilized the parasites at titres of 1:64–1:1024, whereas mucus immobilized parasites at a titre of 1:8–1:16. Areerat[24] produced immunity in catfish with a single exposure to measured numbers of trophozoites and found serum immunoglobulin titres of 1 : 320 (low dose exposure) and 1:5120 (high dose exposure).

Wahli and Meier[25] investigated the properties of serum and mucus of infected and convalescent rainbow trout. They found that during a heavy infection mucus immobilized parasites, whereas serum had a low activity. However, several months post-infection, high serum antiparasite activity was contrasted with a low mucus activity.

Recently, another mechanism of non-specific cellular anti-parasite immunity in fish has been proposed.[26] Non-specific cytotoxic cells (NCC) from the anterior kidney of channel catfish were shown to lyse *T. pyriformis* suggesting that a similar mechanism may occur with *I. multifiliis*. These NCC cells appear to share many characteristics with mammalian natural killer (NK) cells but seem to demonstrate a broader spectrum of activity and may represent an evolutionary precursor of the more specialized, comparable cells in higher vertebrates. Nevertheless, they may serve important functions in the teleost immune response.

Thus, following a controlled infection, fish develop an immune resistance to further infection. Obviously, the use of live parasites to vaccinate fish would constitute a disease risk and safer means have been sought.

Table 18.2 *Protection of channel catfish immunized with ciliary and cellular antigens of* Tetrahymena pyriformis *and* Ichthyophthirius multifiliis *against* I. multifiliis (modified after Goven *et al.*[27]).

Antigen	Dose of antigen (μg)	% Mortality
T. pyriformis cilia	5.0	11.6
Deciliated *T. pyriformis*	10.0	100.0
I. multifiliis cilia	3.5	42.5
Deciliated *I. multifiliis*	2.5–4.5	84.7

Towards a safe vaccine

One basic necessity for producing a vaccine is the ease of production of safe protective antigens. *I. multifiliis* is an obligate pathogen and the production of large amounts of antigen would pose formidable problems. However, the possibility of developing a vaccine against 'Ich' became feasible when it was discovered that channel catfish injected with a closely related ciliate, *Tetrahymena pyriformis* were protected against *I. multifiliis*.[27] *T. pyriformis* is a free-living ciliate and can be cultured in large quantities. When fingerlings were injected with cilia or deciliated cells from each of these organisms and then challenged by co-habitation with infected fish, those immunized with cilia of *T. pyriformis* showed an 11.6 per cent mortality compared to a 42.5 per cent mortality in those vaccinated with cilia from *I. multifiliis* (Table 18.2). Deciliated vaccine preparations gave little or no protection. Further work[28,29] using different doses of *T. pyriformis* cilia (10–25 μg of ciliary protein) confirmed these earlier findings.

The efficacy of vaccination by different routes has also been investigated. Pyle[30] vaccinated channel catfish with *T. pyriformis* cilia by oral, topical application and intramuscular routes. However, only the intramuscularly injected material was immunogenic with serum antibody titres persisting for at least 12 weeks. In juvenile carp injected intraperitoneally with whole, live *T. pyriformis*, an enhanced antibody response occurred following a second injection, again with serum titres persisting for at least 12 weeks.[17] Wolf and Markiw[31] using live *T. thermopila* ($1–2 \times 10^5$ whole cells/ml) or sheared cilia from a similar number of cells administered by a 4-h bath, were able to protect rainbow trout against 'Ich'. The degree of protection obtained, which first developed 4 weeks post-immunization, increased to a maximum 10 weeks after exposure. 'Ich' immunized fish also appeared to be more resistant to natural infections of *Ichthyobodo necator* (costia) especially those fish which were immunized with whole cells. This suggests that immunization against *I. multifiliis* also provides protection against *I. necator*. If the immunization procedure stimulates defences such as NCC, it is possible protection may also be induced against other parasitic ciliates, e.g. *Trichodina* and *Chilodinella*.

Although the protective mechanism is as yet unknown, Goven *et al.*[29] suggested that antibody secreted in the cutaneous mucus caused immobilization of the theronts. Antigenic cross-reactivity has been demonstrated between *T. pyriformis* and *I. multifiliis* by using *in vitro* serological tests.[32] An indirect immunofluorescence test suggested that the cross reactivity was localized in the cilia and pellicle. However, protection to 'Ich' may depend on the strain or isolate of *T. pyriformis* used. Dickerson *et al.*[33] investigated three different isolates of *T. pyriformis* and used alternative methods of cilia preparation. Although the latter did not appear to have any significant effect on induction of protection, there was considerable variation in the degree of protection elicited by the different isolates. Furthermore, Houghton[17] was unable to induce protection in juvenile carp injected intraperitoneally with whole live organisms and challenged with quantified numbers of theronts. Despite the production of serum antibody to *T. pyriformis*, no antigenic cross reactivity was demonstrated to *I. multifiliis* using *in vitro* serological tests.

Future developments

Practical applications of vaccines composed of cilia of either *T. pyriformis* or *T. thermophila* appear promising. *Tetrahymena* ciliate antigens have some distinct advantages; the organism is non-pathogenic and can be grown easily *in vitro* at least on a small scale. However, in the development of any vaccine, it is essential not only to identify the major antigens, but also to reclassify the present strains of *Tetrahymena* spp. and identify the strong immunogens which cross-react with *I. multifiliis*. A recent significant development is the production of a panel of murine monoclonal antibodies to the theronts of *I. multifiliis*[34] which may be of significant value in the characterization of *I. multifiliis* antigens necessary in vaccine production. Obviously, comprehensive trials would be necessary to assess the real field value of a vaccine. As yet no definitive work has been completed on either the optimum temperature at which to immunize, or indeed the longevity of the protection afforded. Wolf and Markiw[31] have suggested that when immunized fish are exposed to natural infections of *I. multifiliis*, they may develop a secondary response which would further enhance protection. The mode of administration is also important. Bath vaccination is the only commercially viable procedure, and while there is conflicting evidence that immersion techniques are effective it would seem likely that topical application of a vaccine which stimulated a secretory immune response may be the most valuable in protection against surface or ectoparasites. Houghton and Matthews[14] have shown that corticosteroids have an immunosuppressant effect on theront-induced protection; thus any factors in intensive culture systems leading to severe stress and, hence, elevated corticosteroid levels could interfere with protective immunity.

Another factor to be considered important is the possibility of the development of carrier states to *I. multifiliis* as suggested by Hoffman;[35] this could lead to a potential reservoir of parasites leading to outbreaks of infection when the fish are stressed.

COSTIASIS

The bodonid flagellate *Ichtyobodo necator* (Henneguy, 1883) (= *Costia necatrix*) is a significant ectoparasite of a variety of freshwater fish[36] which causes skin and gill lesions and mortalities particularly in intensive freshwater salmonid culture systems. It is also now established as an important parasite in marine culture of salmonids[37] and has also been isolated from 'natural' populations of marine flatfish.[38] The occurrence of an acquired immunity to *I. necator* has been demonstrated by Tavolga and Nigrelli[39] and Robertson[40] found a marked periodicity of the parasite on salmonid fry, with a rapid increase in numbers of parasites to a peak at 4 weeks, followed by a significant decline which may be related to an innate or acquired immunity. Chemotherapeutic treatment and other control measures are very limited and detailed studies on the nature of the acquired immunity with a view to potential development of a vaccine would be most valuable in salmonid culture.

TRYPANOSOMIASIS

A number of species of the genera *Trypanosoma* and *Trypanoplasma* (= *Cryptobia*) are quite commonly found in the blood of both freshwater and marine fishes. Although, unlike other flagellates, they are not generally considered to be important pathogens of fish, they have been shown to cause mortality in salmonids and serious problems in young carp. They have also received attention as parasites of fish because of the devastating effects of related species to their mammalian hosts, where their ability to constantly change their antigenic determinants is well known. *Trypanoplasma salmo-*

sitica undergoes a similar antigenic variation in trout.[41,42] Woo[43] demonstrated that goldfish which survived inoculations of either a high or low dose of *T. danilewskyi* all survived a high challenge infection. A greater percentage survival occurred after inoculation with a low dose, suggesting the immune system had time to respond and eliminate the parasite. Passive protection was also conferred with immune plasma, demonstrating the effectiveness of the humoral response. Burreson and Frizzell[44] immunized summer flounder with formalin-killed *T. bullocki* at 20°C and 14°C. Following live challenge, increases in the antibody titre were related to the temperature. However, titre was not necessarily related to protection against the parasite and at low temperatures, high titres of antibody were not able to regulate the presence of the parasite, suggesting cellular mechanisms are also involved in protection.

DIPLOSTOMIASIS

The species of eye-fluke, *Diplostomum* spp. are a serious fishery management problem in many areas of Europe, the USA and Africa. Diplostomiasis can also be a problem in freshwater salmonid culture where aquatic vegetation allows the survival of the intermediate snail hosts. Members of the family Salmonidae appear to be very susceptible to this helminth. Fish are infected by motile cercariae penetrating their integument causing severe damage both along the migration route and in the target organ, the eye. Parasitization of the eye causes cataracts leading to partial or total blindness, abnormal feeding, reduced growth and mortality. Chemical or physical control of the snail hosts of *D. spathaceum* is difficult, although Stables and Chappell[45] have made some limited advances using molluscicides in rainbow trout farms. However, no really effective method of preventing or controlling the disease has been developed.

The limited information on the immune response of fish to infection with eye-fluke has recently been reviewed.[17,46] A problem for vaccine development is the fact that the eye lens is an immunologically privileged site, i.e. the ingress of immunoglobulin or leucocytes is normally prevented. Furthermore, the digenea have developed sophisticated mechanisms to evade the hosts' immune response. Nevertheless, recent work has suggested that vaccination of fish using antigenic preparations of metacercariae may be feasible. Speed and Pauley[47] found that rainbow trout injected with whole or sonicated cercariae or metacercariae with adjuvant (FCA) increased survival time over non-immunized controls following challenge with live cercariae, but no serum antibody could be detected. Survival times increased as the dose of antigen and fish size increased. Using the more sensitive ELISA technique, Bortz *et al.*[48] were able to detect the presence of circulating antibody (reaching maximum titres 3 weeks after immunization) to metacercarial antigens following experimental immunization of rainbow trout with sonicated metacercariae (10–100/fish) and in fish naturally infected. An enhanced secondary response occurred following a booster immunization (50 metacercariae/fish).

Stables and Chappell,[46] using a variety of techniques, were unable to detect humoral antibody in rainbow trout following immunization with suspensions of killed cercariae (without adjuvant) or live infections. They did, however, find a significant decrease in the level of infection in immunized fish but were unable to characterize the mechanism. These promising preliminary results need to be enhanced by more careful purification of cercarial and metacercarial antigens, a thorough analysis of the value of the adjuvant and improved cultivation techniques for the *in vitro* maintenance of the parasite before an economically viable vaccine can be envisaged.

OTHER DISEASES

The development of acquired immunity in fish is known, not only to ciliate and flagellate parasites, but also to other groups, particularly micro- and myxosporideans which are often important parasites of both freshwater and marine fish. *Glugea stephani* (microsporidean) infects flatfish; *Myxosoma cerebralis* and PKX (the causative agent of proliferative kidney disease) are both myxosporideans important in rainbow trout culture.[17]

While fish surviving infections of these parasites develop resistance to further infections, little is understood of the nature or mechanisms of the immunity. Once further information becomes available, the potential for vaccine development may be assessed.

REFERENCES

1 Kennedy, C. R. (1977). The regulation of fish parasite populations. In: *Regulation of Parasite Populations*, ed. G. W. Esch, London: Academic Press, pp. 66–109.
2 Lom, J. (1969). Cold-blooded vertebrate immunity to Protozoa. In: *Immunity to Parasitic Animals*, ed. J. A. Jackson, New York: Appleton-Century-Crofts, pp. 249–265.
3 Snieszko, S. F. (1969). Cold-blooded vertebrate immunity to Metazoa. In: *Immunity to Parasitic Animals*, ed. J. A. Jackson, New York: Appleton-Century-Crofts, pp. 267–275.
4 Kennedy, C. R. (1975). *Ecological Animal Parasitology*, Oxford and Edinburgh: Blackwell.
5 Van Muiswinkel, W. B. and Jagt, L. P. (1984). Host–parasite relationships in fish and other ectothermic vertebrates. *Dev. Comp. Immunol.*, **Suppl. 3,** 205–208.

6 Nigrelli, R. F., Pokorny, K. S. and Ruggieri, G. D. (1976). Notes on *Ichthyophthirius.* A ciliate parasitic on freshwater fishes with some remarks on possible physiological races and species. *Trans. Amer. microsc. Soc.* **95**, 607–613.
7 Dashu, N. and Lien-Siang (1960). Studies on the morphology and life-cycle of *Ichthyophthirius multifiliis* and its control with a description of a new species. *Acta hydrobiol. Sin.,* **2**, 213–315.
8 Ventura, M. T. and Paperna, I. (1985). Histopathology of *Ichthyophthirius multifiliis* infections in fishes. *J. Fish Biol.,* **27**, 185–203.
9 Hines, R. S. and Spira, D. T. (1974). Ichthyophthiriasis in the mirror carp, *Cyprinus carpio* (L). III. Pathology. *J. Fish Biol.,* **6**, 189–196.
10 Hines, R. S. and Spira, D. T. (1974). Ichthyophthiriasis in the mirror carp, *Cyprinus carpio* (L). IV. Physiological dysfunction. *J. Fish Biol.,* **6**, 365–371.
11 Leteux, F. and Meyer, F. P. (1972). Mixtures of malachite green and formalin for controlling *Ichthyophthirius* and other protozoan parasites of fish. *Prog. Fish Cult.,* **34**, 21–26.
12 Van Djuin, C. Jr. (1973). *Diseases of Fishes.* London: Iliffe.
13 Farley, D. G. and Heckmann, R. (1980). Attempts to control *Ichthyophthirius multifiliis* Fouquet (Ciliophora: *Ophryoglenidae*) by chemotherapy and electrotherapy. *J. Fish Dis.,* **3**, 203–212.
14 Houghton, G. and Matthews, R. A. (1986). Immunosuppression of carp (*Cyprinus carpio* L) to ichthyophthiriasis using the corticosteroid triamcinolone acetonide. *Vet. Immunol. Immunopathol.,* **12**, 413–419.
15 Wahli, T., Steiff, K. and Meier, W. (1985). Influence of ascorbic acid on *Ichthyophthirius multifiliis* infections in trout. *Bull. Eur. Ass. Fish Pathol.,* **5**, 86–87.
16 Hines, R. S. and Spira, D. T. (1974). Ichthyophthiriasis in the mirror carp, *Cyprinus carpio* (L). V. Acquired immunity. *J. Fish Biol.,* **6**, 373–378.
17 Houghton, G. (1987). The immune response of carp (*Cyprinus carpio* L) to *Ichthyophthirius multifiliis* (Fouquet, 1876). PhD thesis, Plymouth Polytechnic, Plymouth, England.
18 McCallum, H. I. (1986). Acquired resistance of black mollies, *Peocilia latipinna* to infection by *Ichthyophthirius multifiliis. Parasitol.,* **93**, 251–261.
19 Lahav, M. and Sarig, S. (1973). Observation of laboratory infection of carp by *Ichthyophthirius multifiliis* Fouquet. *Bamidgeh,* **25**, 3–9.
20 Valtonen, E. T. and Keranen, A. L. (1981). Ichthyophthiriasis of Atlantic salmon, *Salmo salar* L at the Montta Hatchery in northern Finland in 1978–1979. *J. Fish Dis.,* **4**, 405–411.
21 Subasinghe, R. P. (1982). Investigation of latent infections of *Ichthyophthirius multifiliis* (Fouquet) in juvenile *Cyprinus carpio.* MSc thesis, Plymouth Polytechnic, Plymouth, England.
22 Beckert, H. (1975). Observations on the biology of *Ichthyophthirius multifiliis* (Fouquet, 1876). PhD thesis, University of South Western Louisiana, Louisiana, USA.
23 Hines, R. S. and Spira, D. T. (1973). Acquired immunity of the mirror carp (*Cyprinus carpio* L) to Ichthyophthiriasis. *Refuah. vet.,* **30**, 17–19.
24 Areerat, S. (1974). The immune response of the channel catfish, *Ictalurus punctatus* (Rafinesque) to *Ichthyophthirius multifiliis.* MS thesis, Auburn University, Alabama, USA.
25 Wahli, T. and Meier, W. (1985). Ichthyophthiriasis in trout: investigation of natural defence mechanisms. In: *Fish and Shellfish Pathology,* ed. A. E. Ellis, London: Academic Press, pp. 347–352.

26 Graves, S. S., Evans, D. L. and Dawe, D. L. (1985). Antiprotozoan activity of non-specific cytotoxic cells (NCC) from the channel catfish (*Ictalurus punctatus*). *J. Immunol.*, **134**, 78–85.
27 Goven, B. A., Dawe, D. I. and Gratzek, J. B. (1980). Protection of channel catfish, *Ictalurus punctatus* Rafinesque, against *Ichthyophthirius multifiliis* Fouquet by immunization. *J. Fish Biol.*, **17**, 311–316.
28 Goven, B. A., Dawe, D. L. and Gratzek, J. B. (1980). Antiprotozoan immunization studies in the channel catfish, *Ictalurus punctatus*. *Proc. World Maricul. Soc.*, **11**, 275–280.
29 Goven, B. A., Dawe, D. L. and Gratzek, J. B. (1981). Protection of channel catfish (*Ictalurus punctatus*) against *Ichthyophthirius multifiliis* (Fouquet) by immunization with varying doses of *Tetrahymena pyriformis* (Lwoff) cilia. *Aquaculture*, **23**, 269–273.
30 Pyle, S. W. (1983). Antigenic and serologic relationships between *Ichthyophthirius multifiliis* Fouquet and *Tetrahymena pyriformis* Lwoff. PhD thesis, University of Georgia, USA.
31 Wolf, K. and Markiw, M. E. (1982). Ichthyophthiriasis: immersion immunization of rainbow trout (*Salmo gairdneri*) using *Tetrahymena thermopila* as a protective immunogen. *Can. J. Fish Aquat. Sci.*, **39**, 1722–1725.
32 Goven, B. A., Dawe, D. L. and Gratzek, J. B. (1981). *In vitro* demonstration of serological cross-reactivity between *Ichthyophthirius multifiliis* Fouquet and *Tetrahymena pyriformis* Lwoff. *Dev. Comp. Immunol.*, **5**, 283–289.
33 Dickerson, H. W., Brown, J., Dawe, D. L. and Gratzek, J. B. (1984). *Tetrahymena pyriformis* as a protective antigen against *Ichthyophthirius multifiliis* infections; comparisons between isolates and ciliary preparations. *J. Fish Biol.*, **24**, 523–528.
34 Dickerson, H. W., Evans, D. L. and Gratzek, J. B. (1986). Production and preliminary characterisation of murine monoclonal antibodies to *Ichthyophthirius multifiliis*, a protozoan parasite of fish. *Am. J. Vet. Res.*, **47**, 2400–2404.
35 Hoffman, G. L. (1976). Protozoan diseases of freshwater fishes: advances and needs. In: *Wildlife Diseases*, ed. L. A. Page, New York and London: Plenum Press, pp. 141–142.
36 Becker, C. D. (1977). Flagellate parasites of fish. In: *Parasitic Protozoa*, Vol. 1, ed. J. P. Kreier, London: Academic Press, pp. 358–412.
37 Ellis, A. E. and Wootten, R. (1978). Costiasis of Atlantic salmon *Salmo salar* L. smolts in sea water. *J. Fish Dis.*, **1**, 389–393.
38 Bullock, A. M. and Robertson, D. A. (1982). A note on the occurrence of *Ichtyobodo necator* (Henneguy, 1883) in a wild population of juvenile plaice, *Pleuronectes platessa* L. *J. Fish Dis.*, **5**, 531–533.
39 Tavolga, W. N. and Nigrelli, R. F. (1947). Studies on *Costia necatrix* Henneguy. *Trans. Amer. Microsc. Soc.*, **66**, 366–378.
40 Robertson, D. A. (1979). Host–parasite interactions between *Ichtyobodo necator* (Henneguy, 1883) and farmed salmonids. *J. Fish Dis.*, **2**, 481–491.
41 Lowe-Jinde, L. (1979). Some observations of rainbow trout, *Salmo gairdneri* Richardson, infected with *Cryptobia salmositica*. *J. Fish Biol.*, **14**, 297–302.
42 Woo, P. T. K., Wehnert, S. D. and Rodgers, D. (1983). The susceptibility of fishes to haemoflagellates at different ambient temperatures. *Parasitol.*, **87**, 385–392.
43 Woo, P. T. K. (1981). Acquired immunity against *Trypanosoma danilewskyi* in goldfish, *Carassius auratus*. *Parasitol.*, **83**, 343–346.

44 Burreson, E. M. and Frizzell, L. T. (1986). The seasonal antibody response of juvenile summer flounder (*Paralichthys dentatus*) to the hemoflagellate *Trypanoplasma bullocki. Vet. Imm. Immpathol.*, **12**, 395–402.
45 Stables, J. N. and Chappell, L. H. (1986). The epidemiology of diplostomiosis in farmed rainbow trout from northeast Scotland. *Parasitol.*, **92**, 699–710.
46 Stables, J. N. and Chappell, L. H. (1986). Putative immune response of rainbow trout, *Salmo gairdneri*, to *Diplostomum spathaceum* infections. *J. Fish Biol.*, **29**, 115–122.
47 Speed, P. and Pauley, G. B. (1985). Feasibility of protecting rainbow trout, *Salmo gairdneri* Richardson, by immunizing against the eye fluke, *Diplostomum spathaceum. J. Fish Biol.*, **26**, 739–744.
48 Bortz, B. M., Kenny, G. E., Pauley, G. B., Garcia-Ortigoza, E. and Anderson, D. P. (1984). The immune response in immunized and naturally infected rainbow trout (*Salmo gairdneri*) to *Diplostomum spatheceum* as detected by enzyme-linked immunosorbent assay (ELISA). *Dev. Comp. Immunol.*, **8**, 813–822.

19

Immune Control of Sexual Maturation in Fish

C. J. Secombes

In aquaculture there is a need to control sexual maturation in fish in order to prevent diversion of somatic growth into the gonads, loss of flesh quality and appearance of commercially undesirable secondary sexual characteristics associated with gonadal development. This is especially true with salmonids, where sexual maturation of precocious parr and grilse can have a severe economic effect by arresting the growth of parr at a non-commercial size and forcing the harvesting of grilse over a short period lowering the market value. There is also a need to inhibit fertility, to prevent escapees from fish farms or stray fish from salmon ranching contributing to natural gene pools. The most effective way to ensure sexual maturation does not occur is to rear sterile fish, yet despite numerous experimental approaches this has still to be achieved for bisexual populations. Immunological methods to control gonadal development and fertility are becoming increasingly effective in mammals[1,2] and their applicability to fish will now be discussed.

There are two types of target for immune intervention, the gametes themselves (and possibly their associated somatic cells) or the hormones which control gametogenesis. These hormones include the sex steroids, the gonadotropin hormones (GTH) and gonadotropin-releasing hormone (GnRH). Antibodies that form complexes with these hormones can theoretically interfere with their biological action (termed immunoneutralization) so inhibiting sexual maturation. For either of the targets two types of immunization strategy can be adopted; active immunization or passive immunization. Active immunization relies on the animal mounting an immune response against the injected antigen. This may require several weeks to appear but once elicited can have a long-lasting inhibitory effect. Immune responses can be induced against 'self' antigens in addition to 'non-self' antigens (bacteria, viruses, etc.) and such reactions are termed autoimmune responses. Passive immunization involves injecting preformed antibodies raised in other animals. This can have an immediate effect, although the effective period is limited to the biological life of the injected antibodies. If

FISH VACCINATION
ISBN 0-12-237485-1

the action of the antibodies is reversible, booster injections will be continually required and this would be unacceptable in terms of the amounts of antiserum required, the repeated handling of fish and the risks of eliciting deleterious reactions against the large quantities of injected foreign proteins. Therefore, active immunization is normally the procedure of choice.

HORMONES

Production of eggs or sperm in the gonads is under the control of GTH released from the pituitary (Fig. 19.1). Two types of GTH have been isolated in most vertebrates, luteinizing hormone (LH) and follicle stimulating hormone (FSH), but in fish the number of GTH is still controversial. One glycoprotein GTH is known to induce gametogenesis and is involved in the initiation of vitellogenin synthesis, oocyte maturation, ovulation and spermiation and is thus functionally similar to both LH and FSH. A second carbohydrate-poor GTH has been isolated from some fish and promotes vitellogenin incorporation into oocytes. In addition to their effects upon gametogenesis GTH stimulate the production of androgens and oestrogens, the sex steroids, from the gonad. These have a role in the reproductive process but also act on tissues not directly involved with reproduction, giving rise to the secondary sexual characteristics. The synthesis and secretion of GTH is itself regulated by the binding of GnRH released from the hypothalamus. This hormone travels to the pituitary in the pituitary portal blood vessels (primitive teleosts) or along nerve fibres terminating in the pituitary (advanced teleosts) (see Ball[3]). In addition, the sex steroids have a negative feedback effect upon GTH secretion.

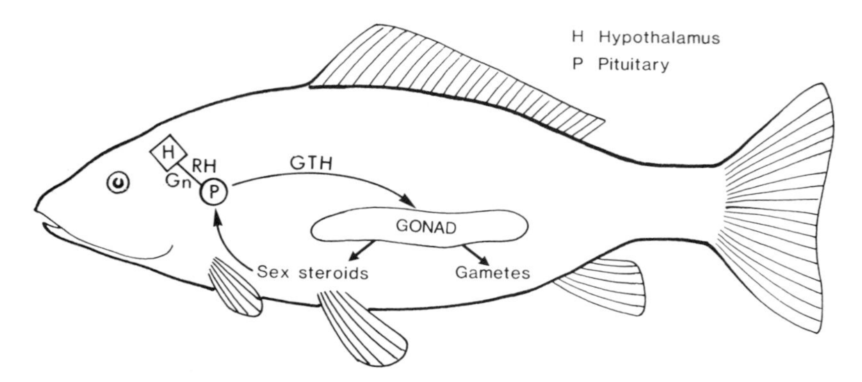

Fig. 19.1 Diagram of the major reproductive hormone pathways.

Sex steroids

Steroid hormone molecules are too small (MW 250–350) to elicit antibody production by themselves but can do so when coupled to larger carrier molecules and in mammals active immunization with such conjugates does induce steroid specific antibodies. However, the resulting inhibition of steroid hormones prevents the normal negative feedback on the pituitary and results in hypersecretion of pituitary GTH, with a positive correlation being found between antibody titre and GTH concentration. The GTH in turn stimulates the gonad and causes increased production of the sex steroids. In males immunized against testosterone, the testis weight actually increases and serum testosterone levels can increase 100 times.[4] In females, immunization against sex steroids leads to similar hormonal changes which increase the ovulation rate and this technique is already being used as a 'fecundity' vaccine for sheep.[5,6] As similar negative feedback effects on GTH secretion appear to operate in fish[7] sex steroids do not appear to be an appropriate choice for immunization.

Gonadotropin hormones

Immunization against GTH in mammals is rather variable but has had some success. Neutralization of LH inhibits sex steroid production, thus preventing ovulation and spermatogenesis and delaying sexual maturation.[8] Neutralization of FSH is not so effective: it can reduce sperm production but has no effect upon testicular endocrine functions or ovarian development. The main drawbacks of immunization against GTH in fish are (a) the limited amounts of purified fish GTH available, (b) the high probability of cross-reactions with other hormones, such as thyroid stimulating hormone (TSH), with important roles in processes other than reproduction and which have considerable parts of their molecular structure in common and (c) the relatively large quantities of hormone to be neutralized which could give rise to dangers associated with immune complex formation (e.g. damage to the kidneys). Nevertheless, passive immunization against GTH has been carried out in fish. Short-term daily administration of excess rabbit antibodies to the glycoprotein GTH in vitellogenic salmon, flounder and catfish results in decreased levels of oestradiol and vitellogenin, decreased yolk incorporation and an inhibition of ovulation;[9,10] but prolonged treatment fails to inhibit rapid ovarian growth.[11] Passive immunization of vitellogenic fish against the carbohydrate-poor GTH also decreases the amount of yolk incorporated into the ovary, causing atresia of yolky oocytes[12] but has none of the other effects.[10] Multiple immunization of immature fish with antibodies to the glycoprotein GTH over a 1-month period does inhibit ovarian growth, as evidenced by a lower gonadosomatic index (GSI) and impeded oocyte development.[13] Whether long-term inhibition of gonad development can be achieved by active immunization of immature fish against the glycoprotein GTH is not known.

Gonadotropin-releasing hormone

The pituitary gonadotropes are regulated by the hypothalamus and require a continual input of the hypothalamic neurohormone GnRH. GnRH is a decapeptide and like the sex steroids it is poorly immunogenic unless coupled to a larger carrier molecule. An advantage of GnRH, however, is that the amount of hormone to be neutralized is not very large and consequently the amount of antibody required is modest and attainable. In mammals immunization against GnRH establishes a selective barrier between the hypothalamus and the anterior pituitary and in the absence of GnRH, GTH secretion is reduced and pituitary stores of GTH are depleted. This causes failure of the gonad to develop in young animals or involution of testicular size, cessation of spermatogenesis and severe reduction of testosterone secretion in adult males and cessation of ovulation and considerable lowering of oestrogen secretion in adult females.[14,15] Thus the available data support the effectiveness of immunization against GnRH for chronically suppressing reproductive activity in mammals. Recently, the primary structure of salmon GnRH has been successfully determined[16] and differs from mammalian GnRH by only two amino acids. Whether anti-GnRH antibodies can be elicited in fish remains to be investigated.

GAMETES

The gametes are derived from precursor germ cells in the testis and ovary by spermatogenesis and oogenesis. In most teleosts the testis is similar to that of mammals and contains, in immature fish, spermatogonia, which increase in number by mitosis, along the entire length of the seminiferous tubules. During spermatogenesis they undergo two meiotic divisions, when they are termed primary and secondary spermatocytes. This serves to reduce the chromosome number in the resulting spermatids and spermatozoa. During these divisions the germ cells are surrounded by Sertoli cells and only when the Sertoli cell processes separate are the haploid cells released into the tubule lumen. Leydig cells are also present in the testis and are involved in steroidogenesis. In the ovary mitotic proliferation of oogonia similarly precedes oogenesis, when oogonia undergo two meiotic divisions. However, oogenesis is arrested during the first meiotic division while enormous growth of the oocytes occurs and is only completed relatively late under appropriate hormonal control. Seasonal spawners undergo these changes in the gonad annually.

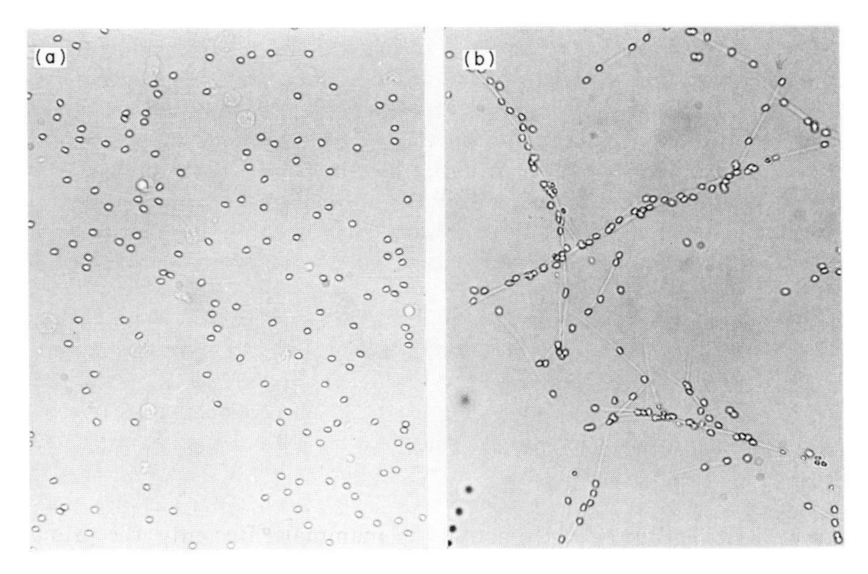

Fig. 19.2 Agglutination of spermatozoa after incubation with (a) normal serum or (b) autoimmune serum. × 450.

ACTIVE AUTOIMMUNIZATION

Mature fish

Spermatozoa appear relatively late during ontogeny, are shielded from the immune system by a Sertoli cell (blood–testis) barrier and are amongst the most autoantigenic cells known in mammals (Hogarth, 1982).[17] It is not surprising, therefore, that injection of mature rainbow trout with an homogenate of mature testis (consisting almost entirely of spermatozoa) emulsified in adjuvant also elicits an autoimmune response.[18] This response is testis specific and is characterized by the appearance of sperm agglutinating autoantibodies (Fig. 19.2) in the serum[19] and macrophage invasion in the seminiferous tubules. The macrophage response is focal (Fig. 19.3), with some areas of the testis remaining normal. Following ingestion of spermatozoa the macrophages collect to form severe granulomatous lesions within the tubules. Such lesions culminate in complex giant cell granulomatous lesions, with all the spermatozoa phagocytosed and the original tubule structure no longer recognizable. Immunoglobulin (Ig) can be detected both within the seminiferous tubules undergoing macrophage invasion and within milt samples from these fish. As Ig is normally absent from these sites this suggests that a blood–testis barrier to serum proteins is broken down during autoimmune orchitis. The lesions are unlikely to be solely antibody-induced inflammatory reactions though, because there is no correlation between the amount of antibody present and the degree of pathological damage in the testis. Thus, some fish with high antibody titres show no testicular lesions and fish with no or low titres range from no reaction to severe lesions.

In contrast to testis, no effects have been seen in ovaries from fish immunized with mature testis or oocyte membrane homogenates in adjuvant.

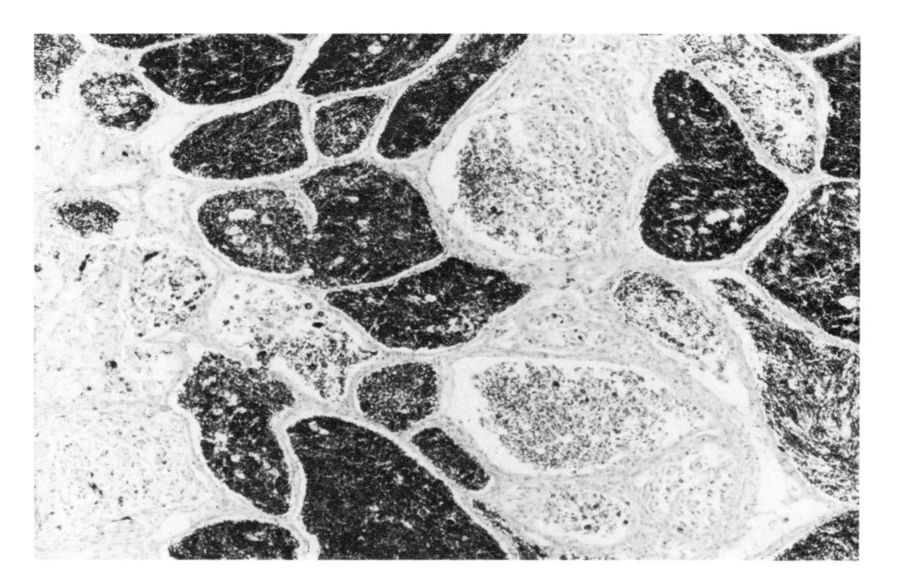

Fig. 19.3 Focal macrophage invasion of the testis from an autoimmunized rainbow trout. × 180.

Immature fish

Autoimmune reactions can clearly be elicited against the testis of mature fish but it is important to know whether fish can be prevented from maturing after administration of the gonad vaccine. Studies using immature trout injected with an homogenate of mature testis in adjuvant have revealed that although antibodies are produced there is no damage to the testes.[18] Thus the autoantigens present on spermatozoa are not present on spermatogonia. Experiments carried out to determine at what stage during spermatogenesis spermatozoal autoantigens do appear have found them to be present on spermatozoa and spermatids but not spermatocytes.[20,21] Thus, experimental orchitis in maturing fish can only be elicited if the testes contain spermatozoa and spermatids. No other cell types appear to be involved in the antibody or macrophage response. This did not exclude the possibility that autoantigens not present on spermatozoa might occur on the other germ cell stages. However, injection of maturing fish with testis material containing all germ stages from spermatogonia to spermatozoa also failed to reveal the presence of distinct autoantigens on earlier germ cell stages. In fish the blood–testis barrier appears to develop after meiosis and therefore only spermatozoa and spermatids can be considered to be protected by it.[22] This agrees with the observation that only spermatozoa and spermatids are autoantigenic and would need to be shielded from the immune system.

From a practical viewpoint the late appearance of autoantigens on postmeiotic germ cells means that autoimmunity can only be elicited relatively late during spermatogenesis and that sexual maturation is not blocked by prior immunization with testis/adjuvant. Indeed even the extensive granulomatous tissue formed within seminiferous tubules during the autoimmune response in mature fish did not prevent subsequent maturations.[23] Thus the development of secondary sexual characteristics and increase in gonad size (GSI) associated with sexual maturation is not prevented using this approach. It may have some application in providing functionally sterile fish, as the sperm produced are often prevented from being released by the long-term granulomas within the sperm duct.

PASSIVE IMMUNIZATION

Antibodies against germ cells can be used to kill their target cells directly or indirectly following the induction of autoimmune reactions. The latter approach, using fish anti-sperm antibodies, has not received much attention as a practical means to prevent sexual maturation because prior immunization with adjuvant is required and like active immunization it can only elicit autoimmune orchitis in fish that are already mature.[24] Therefore, the use of antibodies against precursor germ cell antigens and direct killing of targets has more potential, especially when using monoclonal antibodies (MoAb) which can be raised against individual gonad specific antigens.[25]

In vivo killing of germ cells by passively transferred antibodies has still to be achieved but the necessary preliminary experiments look promising. For instance, MoAb raised against carp spermatozoa and selected for their reactivity with antigens common to all sperm cell precursors, oogonia and early prophase oocytes, are able to kill isolated precursor germ cells *in vitro* using complement-mediated cytotoxicity.[25] Unfortunately, this technique requires mammalian complement to effect cytotoxicity and has a number of disadvantages for *in vivo* cytotoxicity. For this reason, toxins coupled to the antibody molecule may ultimately prove more appropriate. Intravascular injection of such MoAb into male fish has shown that IgG MoAb are able to reach and localize on the surface of the designated target cells (precursor germ cells outwith the blood–testis barrier), although IgM MoAb remain predominantly intravascular.[25] This is not the case in females where both types of MoAb can 'home' to the oogonia and early prophase oocytes. Thus, passive immunization with MoAb gives the possibility of killing premeiotic germ cells *in vivo*, particularly spermatogonia and oogonia, as early as possible and before they have increased considerably in number by mitogenesis.

CONCLUSIONS

It is clear that immunological approaches to control maturation in fish do have considerable potential and warrant further investigation as an alternative to the many other approaches such as irradiation, chemosterilization and hormone administration, which have so far proved unreliable, uneconomical or unsuitable for use on fish destined for the table (for reviews of other approaches see Laird *et al.*,[26] Donaldson and Hunter,[27] and Refstie[28]). As with vaccination against disease the major benefit of a gonad vaccine would be in the prevention of sexual maturation rather than its 'cure'. Thus, the inhibition of gonad development by 'prophylactic' immunization is desirable. Active immunization against autoantigens on germ cells does not meet this objective because an autoimmune response is only elicited in fish that are already mature. Whether active immunization against Leydig cells in the testis or germ cells in the ovary, both of which possess autoantigens in mammals, will be of more practical use has yet to be ascertained. Active immunization against the reproductive hormones has more promise but there is little experimental data in fish. One drawback of active immunization is the need for an adjuvant to induce good anti-hormone responses. Freund's complete adjuvant (FCA) is commonly used and is particularly effective in overcoming the state of natural tolerance to many cells and molecules but also induces granulomas and this may preclude its eventual use in vaccines. Some progress is being made in the search for alternative adjuvants and muramyl dipeptides in particular have already been shown to be as good as FCA at eliciting anti-hormone responses in mammals.[29] Passive immunization against hormones is unlikely to be practicable, for reasons discussed earlier, but passive immunization to kill specific cell types may be of future interest. MoAb/toxin conjugates able to 'home' to target cells may prove to be an efficient means to selectively kill germ cells or somatic cells secreting hormones *in vivo*. However, the MoAb would have to be carefully chosen on a range of criteria, such as the ability to traverse blood–tissue barriers and would probably have to be raised against the cells of each species to be sterilized as cross-reactions between species do not appear to allow efficient killing.[25]

REFERENCES

1 Talwar, G. P. (1980). *Immunology of Contraception.* Edward Arnold.

2 Crighton, D. B. (1984). *Immunological Aspects of Reproduction in Mammals.* London: Butterworths.

3 Ball, J. N. (1981). Hypothalamic control of the pars distalis in fishes, amphibians and reptiles. *Gen. Comp. Endocrinol.*, **44**, 135–170.

4 Haynes, N. B. and Southee, J. A. (1984). Effects of immunization against steroid hormones on male endocrinology. In: *Immunological Aspects of Reproduction in Mammals,* ed. D. B. Crighton, London: Butterworths, pp. 427–444.

5 Scaramuzzi, R. J., Davidson, W. G. and Van Look, P. F. A. (1977). Increasing the ovulation rate of sheep by active immunisation against an ovarian steroid, androstenedione. *Nature,* **269**, 817–818.

6 Land, R. B., Morris, B. A., Baxter, G., Fordyce, M. and Forster, J. (1982). Improvement of sheep fecundity by treatment with antisera to gonadal steroids. *J. Reprod. Fertil.*, **66**, 625–634.

7 Billard, R., Richard, M. and Breton, B. (1977). Stimulation of gonadotropin secretion after castration in rainbow trout. *Gen. Comp. Endocrinol.*, **33**, 163–165.

8 Wickings, E. J. and Nieschlag, E. (1984). Effects of immunization against gonadotropins on male reproductive functions. In: *Immunological Aspects of Reproduction in Mammals,* ed. D. B. Crighton, London: Butterworths, pp. 419–426.

9 Nath, P. and Sundararaj, B. I. (1977). Inhibition of vitellogenesis in the catfish *Heteropneustes fossilis* (Bloch) by antiserum raised against partially-purified conspecific gonadotropin. *Anat. Record,* **187**, 773–774.

10 Ng, T. B., Campbell, C. M. and Idler, D. R. (1980). Antibody inhibition of vitellogenesis and oocyte maturation in salmon and flounder. *Gen. Comp. Endocrinol.*, **41**, 233–239.

11 Wiegand, M. D. and Idler, D. R. (1984). Failure of antibody to carbohydrate-rich gonadotropin to inhibit rapid ovarian growth in landlocked Atlantic salmon. *Gen. Comp. Endocrinol.*, **55**, 260–268.

12 Ng, T. B., Idler, D. R. and Burton, M. P. (1980). Effects of teleost gonadotropins and their antibodies on gonadal histology in winter flounder. *Gen. Comp. Endocrinol.*, **42**, 355–364.

13 Wiegand, M. D. and Idler, D. R. (1983). Impairment of early ovarian growth in landlocked Atlantic salmon by an antibody to carbohydrate-rich gonadotropin. *Gen. Comp. Endocrinol.*, **49**, 210–219.

14 Fraser, H. M. (1980). Inhibition of reproductive function by antibodies to luteinizing hormone releasing hormone. In: *Immunological Aspects of Reproduction and Fertility Control,* ed. J. P. Hearn, Lancaster: MTP Press Ltd, pp. 143–171.
15 Schanbacher, B. D. (1984). Active immunization against LH–RH in the male. In: *Immunological Aspects of Reproduction in Mammals,* ed. D. B. Crighton, London: Butterworths, pp. 345–362.
16 Sherwood, N., Eiden, L., Brownstein, M., Spiess, J., Rivier, J. and Vale, W. (1983). Characterization of a teleost gonadotropin-releasing hormone. *Proc. Natl. Acad. Sci. USA,* **80**, 2794–2798.
17 Hogarth, P. J. (1982). *Immunological Aspects of Mammalian Reproduction,* Glasgow: Blackie.
18 Secombes, C. J., Lewis, A. E., Laird, L. M., Needham, E. A. and Priede, I. G. (1985). Experimentally induced immune reactions to gonad in rainbow trout (*Salmo gairdneri*). In: *Fish Immunology,* eds. M. J. Manning and M. F. Tatner, London: Academic Press, pp. 343–355.
19 Secombes, C. J., Lewis, A. E., Laird, L. M., Needham, E. A. and Priede, I. G. (1984). Agglutination of spermatozoa by autoantibodies in the rainbow trout, *Salmo gairdneri*. *J. Fish Biol.,* **25**, 691–696.
20 Secombes, C. J., Lewis, A. E., Needham, E. A., Laird, L. M. and Priede, I. G. (1985). Appearance of autoantigens during gonad maturation in the rainbow trout (*Salmo gairdneri*). *J. Exp. Zool.,* **233**, 425–431.
21 Secombes, C. J., Laird, L. M. and Priede, I. G. (1987). Immunological approaches to control maturation in fish. II. A review of the autoimmune approach. *Aquaculture,* **60**, 287–302.
22 Parmentier, H. K., Boogaart van den, J. G. M. and Timmermans, L. P. M. (1985). Physiological compartmentation in gonadal tissue of the common carp (*Cyprinus carpio* L.). A study with horseradish peroxidase and monoclonal antibodies. *Cell Tissue Res.,* **242**, 75–81.
23 Secombes, C. J., Needham, E. A., Laird, L. M., Lewis, A. E. and Priede, I. G. (1985). The long-term effects of auto-immunologically induced granulomas on the testes of rainbow trout, *Salmo gairdneri* Richardson. *J. Fish Biol.,* **26**, 483–489.
24 Secombes, C. J., Lewis, A. E., Laird, L. M., Needham, E. A. and Priede, I. G. (1985). Role of autoantibodies in the autoimmune response to testis in rainbow trout (*Salmo gairdneri*). *Immunology,* **56**, 409–415.
25 Secombes, C. J., Winkoop van, A., Boogaart van den, J. G. M., Timmermans, L. P. M. and Priede, I. G. (1986). Immunological approaches to control maturation in fish. I. Cytotoxic reactions against germ cells using monoclonal antibodies. *Aquaculture,* **52**, 125–135.
26 Laird, L. M., Wilson, A. R. and Holliday, F. G. T. (1980). Field trials of a method of induction of autoimmune gonad rejection in Atlantic salmon (*Salmo salar* L.). *Reprod. Nutr. Develop.,* **20**, 1781–1788.
27 Donaldson, E. M. and Hunter, G. A. (1982). Sex control in fish with particular reference to salmonids. *Can. J. Aquat. Sci.,* **39**, 99–110.
28 Refstie, T. (1982). Practical application of sex manipulation. In: *Reproductive Physiology of Fish,* eds. C. J. J. Richter and H. J. Th. Goos, Pudoc, Wageningen, pp. 73–77.

29 Carelli, C., Ralamboranto, L., Audibert, F., Gaillard, J., Briquelet, N., Dray, F., Fafeur, V., Haour, F. and Chedid, L. (1985). Immunological castration by a totally synthetic vaccine: modification of biological properties of LH–RH after conjugation to adjuvant-active muramyl peptide. *Int. J. Immunopharmac.*, **7**, 215–224.

Glossary

active immunization. Stimulation of an individual's immune responses in order to confer protection against disease. Effected by exposure to protective antigens either during an infection or by vaccination.
adjuvant. Substance administered with antigen which non-specifically enhances the immune response.
allograft. Graft exchanged between two genetically dissimilar individuals of the same species.
antibody. Protein with the molecular properties of an immunoglobulin and capable of binding with antigen.
antigen. A substance which binds with antibody. A molecule (protein or polysaccharide) which elicits a specific immune response (*see* immunogen).
bacterin. A vaccine consisting of a suspension of bacterial cells that have been killed by chemical or physical means.
BSA. Bovine serum albumin, an antigen often used experimentally.
carrier. An individual who harbours a particular pathogenic microorganism but who shows no clinical signs of disease and who is potentially able to transmit the pathogen to others.
cell-mediated immunity (CMI). Specific immunity which is dependent upon the presence of T-lymphocytes and responsible for reactions such as allograft rejection, mixed leucocyte reactions (MLR). The effector cells may be other than a lymphocyte.
cfu. Colony-forming units. A measure of counting viable bacterial cells.
ellipsoid. A sheath of phagocytic cells surrounding arterioles and capillaries in the spleen.
ELISA. Enzyme-linked immunosorbant assay.
epitope. A specific site on a molecule with antigenic properties.
haemolysin. A substance which lyses erythrocytes.
haematopoietic. Blood-forming.
HGG. Human gamma globulin. Human immunoglobulin, often used as an experimental antigen.
humoral immunity. Specific immunity mediated by antibodies.
immunogen. A substance which elicits humoral or cell-mediated immunity.
immunoglobulin. Another name for antibody.
immunopotentiation. Artificial augmentation of the immune response in a general sense. Produced by a wide range of agents e.g. adjuvants.

immune tolerance. An immunological response consisting of the development of specific non-reactivity to an antigen capable, in other circumstances, of inducing cell-mediated or humoral immunity.
interferon. Group of proteins with antiviral activity.
lipopolysaccharide. An antigenic component of the wall of Gram-negative bacteria. Contains O-antigen (endotoxin).
mixed leucocyte reaction (MLR). The proliferation of lymphocytes which results when lymphocytes from two individuals are cultured together for 3–5 days.
monoclonal antibodies. Antibodies specific for a single epitope on an antigen.
ontogeny. The sequence of development of an individual from the fertilized ovum to the mature adult.
passive immunization. The use of antibodies from an immune individual to confer short-term immunity in a non-immune individual.
pathogenic factors. Factors produced by a pathogen which induce a pathological effect in the host.
pfu. Plaque forming units. A measure of virus particles assessed by counting the number of plaques (areas of cell death) in a monolayer of tissue culture cells.
pheromone. A chemical, released by an animal into its surroundings, which influences the behaviour or physiology of other individuals of the same species.
pili. Also known as fimbriae. Filamentous appendages which project from the surface of certain Gram-negative bacteria.
protective antigens. Those antigens of a pathogenic microorganism which, if given alone, will stimulate an immune response capable of providing protection against that organism.
serotype. Subtype within a bacterial or viral species identified by serological methods i.e. by detecting differences in the surface antigens of microorganisms on addition of specific antibody.
$TCID_{50}$. (Tissue culture infective dose). That dilution of a viral suspension which causes infection in 50% of tissue culture cell preparations. It is a measure of the number of infective viral particles present.
T-dependent antigen. (Thymus-dependent antigen). An antigen that does not stimulate an antibody response unless lymphocytes derived from the thymus are present (called T-helper lymphocytes). Cooperation between T-helper and B-lymphocytes is required for the latter to respond to such antigens by maturation into antibody-producing cells.
virulence factors. Factors produced by a pathogen which aid its survival within the host and cause disease.

Index